HARDWARE DESIGN AND PETRI NETS

T0140587

Hardware Design
and Petri Nets

Edited by

Alex Yakovlev
University of Newcastle upon Tyne

Luis Gomes
Universidade Nova de Lisboa

and

Luciano Lavagno
Universitá di Udine

KLUWER ACADEMIC PUBLISHERS
BOSTON / DORDRECHT / LONDON

A C.I.P. Catalogue record for this book is available from the Library of Congress.

ISBN 978-1-4419-4969-1

Published by Kluwer Academic Publishers,
P.O. Box 17, 3300 AA Dordrecht, The Netherlands.

Sold and distributed in North, Central and South America
by Kluwer Academic Publishers,
101 Philip Drive, Norwell, MA 02061, U.S.A.

In all other countries, sold and distributed
by Kluwer Academic Publishers,
P.O. Box 322, 3300 AH Dordrecht, The Netherlands.

This printing is a digital duplication of the original edition.

Printed on acid-free paper

Second Printing 2003.

All Rights Reserved
© 2000 Kluwer Academic Publishers, Boston
Softcover reprint of the hardcover 1st edition 2000
No part of the material protected by this copyright notice may be reproduced or
utilized in any form or by any means, electronic or mechanical,
including photocopying, recording or by any information storage and
retrieval system, without written permission from the copyright owner.

Contents

Preface

With the advent of sub-micron VLSI technology, which will soon enable a billion of transistors to be placed on a single chip, hardware design becomes a big challenge. Future VLSI circuits will often be systems-on-chip, whose subsystems include processors, memory banks and input/output controllers. Problems with distributing the global clock between these subsystems, treated as Intellectual Property (IP) cores, are unavoidable. Systems-on-chip will effectively lose the *global notion* of physical time and permit actions in different parts of the systems to be executed in parallel or independently of one another. Such hardware systems will inevitably become more *asynchronous* and *concurrent*, thus causing a sharp increase in complexity in their dynamic behaviour. To cope with the growing complexity and with the need to meet time-to-market demands, designers will require adequate development techniques and tools to enable the analysis, synthesis and verification of concurrent hardware.

The cornerstone of a viable hardware design technology is a good language for adequate modelling and specification of system behaviour. In the past, when systems were predominantly synchronous and sequential, such a language was provided by the model of a Finite State Machine (FSM). As systems become more asynchronous and concurrent, an FSM is inadequate because it is based on states and transitions, inherently sequential and global concepts. Even if the FSM representation allows modelling of non-sequential effects in hardware such as races, the model still uses the notion of a global state, and any concurrent transitions are modelled as a set of possible interleavings of state transitions. Use of an FSM composition may only offer a partial (though often successful!) solution to the modelling and verifying of concurrent systems because it requires constructing a product FSM, which, apart from combinatorial state explosion, faces the problem of adequate interpretation of concurrency and synchronisation between transitions in different FSMs.

Petri nets can act as FSMs if the modelled system is totally sequential. However, if there is an explicit need to model concurrency without showing it in its interleaving form, even for the purposes of a more compact representation, Petri nets have proved most adequate for doing so.

The area of hardware design has traditionally been a fertile field for research in concurrency and Petri nets. Many new ideas about modelling and analysis of concurrent systems, and Petri nets in particular, originated in theory of (asynchronous) digital circuits. For example, the theory of speed-independent circuits by D.E. Muller and W.C. Bartky laid a foundation to the important concepts of feasible sequences, cumulative states, finite equivalence classes, confluence, semi-modularity and so on. Similarly, the theory and practice of digital circuit design have always recognised Petri nets as a powerful and easy-to-understand modelling tool.

The history of the "marriage" between digital hardware and Petri nets has witnessed many bright moments. One such example is the seminal work on parallel computers, asynchronous structures and Petri nets carried out at MIT in the seventies, which involved J.B. Dennis, F. Furtek, M. Hack, J.R. Jump, D. Misunas, S. Patil, P.S. Thiagarajan and others. Another example is the increase of interest to Petri nets and their interpretations in the nineties, driven by the Design Automation community. Signal Transition Graphs (STGs) and software packages built around this model (SIS, ASSASSIN, Petrify) have made Petri nets a practical tool in the hands of VLSI designers. In the last few years, parts of industrial strength microprocessors and interfaces have been designed using STGs, and this process is on its rise. The ever-growing demand for design automation in the IT sector, to build various types of computer-based systems, creates many opportunities for Petri nets to establish their role of a formal backbone in future tools for constructing systems that are distributed, concurrent and asynchronous. Petri nets have already proved very effective in supporting algorithms for solving key problems in synthesis of hardware control circuits. However, since the front end to any realistic design flow in the future is likely to rely on more pragmatic Hardware Description Languages (HDLs), such as VHDL and Verilog, it is crucial that Petri nets are well interfaced to such languages.

The papers collected in this book cover the scope of applications of Petri nets in hardware design, as it is seen at present, sufficiently well. The material is split into five parts, which cover aspects of behavioural modelling, analysis and verification, synthesis from Petri nets and STGs, design environments based on high level Petri nets and HDLs and finally performance analysis using Petri nets.

The three chapters in Part I (Hardware Modelling using Petri Nets) illustrate how Petri nets can be used to model the behaviour of hardware at different levels of functional abstraction. The first chapter, written by R. Wollowski and J. Beister, extends the current status of circuit modelling at a very fundamental level, with a number of new notions of causality, exhibited by circuits (possibly with analogue components) captured in a true concurrency framework. Then, F. Xia and I. Clark use Petri nets to formalise the modelling and to facilitate the mechanical analysis of a non-trivial asynchronous communication mechanism, previously dealt with in a much less automatable way. D.H. Schaefer and J.A. Sosa present, in the third chapter, a large set of modelling constructs or operators, based on high-level Coloured Petri nets, that covers many types of components found in computer-based systems, starting from elementary gates and leading to massively parallel processors. This also offers a unified and intuitive way of explaining the behaviour of modern computing and communication systems.

Part II (Model Analysis and Verification for Asynchronous Design) discovers new solutions in the area of analysis of the two popular specification models of circuits, STGs and Change Diagrams. Change Diagrams, although relatively restrictive (they don't capture choice), can model systems with the AND and OR forms of causality. In chapter 4, U. Schwiegelshohn and L. Thiele, for the first time, study this model as a class of dynamic min-max graphs. They provide efficient algorithms for analysis of both untimed and timed Change Diagrams, with respect to liveness and boundedness. The fifth chapter, written by R. Meyer and P.S. Thiagarajan, addresses the problem of model checking for a subclass of choice-free STGs based on the linear time temporal logic LTrL. Chapter 6, written by A. Kovalyov, improves on the structural approach to the computation of concurrency relation, one of the key problems in STG analysis and STG-based circuit synthesis. The new polynomial algorithm lifts the previous restriction on the class of Petri nets, Free-Choice, to that of Regularity, thus moving closer to the class of STGs that are synthesable by exponential, state-based, algorithms.

Part III (Theory and Practice of Petri net based Synthesis) looks at various issues involved in synthesis of digital hardware using Petri nets and STGs. Chapter 7, written by N. Marranghello, J. Mirkowski and K. Bilinski, provides an overview of methods used to synthesise synchronous digital systems specified by Petri nets. I. Blunno and L. Lavagno, in chapter 8, describe an automated method for deriving STGs from behavioral Verilog HDL for subsequent implementation into an asynchronous control circuit by an automatic synthesis tool (Petrify). This approach shows how Petri nets and asynchronous synthesis tools can be placed into the overall asynchronous design flow that is maximally oriented to a designer who is non-expert in Petri nets or STGs, and has limited experience in asynchronous design. Although the paper shows only the first idea of this approach it is believed to be the most natural way of adapting Petri nets in an industrial CAD environment. Finally, in the last chapter of this part, S.-H.Chung and S.B. Furber, present the design of control circuits for an industrial strength microprocessor Amulet 3; namely, they illustrate how an asynchronous Instruction Prefetch Unit is constructed using STGs and tool Petrify.

Part IV (Hardware Design Methods and Tools) covers aspects of designing systems, that are either hardware or hardware-software controllers, from (high-level) Petri nets specifications maximally using the standard design and simulation environments. Chapter 10, written by P. Rokyta, W. Fengler, and Th. Hummel, describes a concept of the automated design flow, starting from a specification in coloured Petri nets, unfolded to Hardware Petri net description, which is subsequently converted to fully synthesisable (into synchronous logic) VHDL. In chapter 11, R.J. Machado, J. M. Fernandes, A. J. Esteves, and H.D. Santos, present an evolutionary approach to the use of Petri net based models in designing (again, synchronous) systems of increasing complexity from parallel controllers to hardware-software codesign. Finally, D. Protheroe describes in a very systematic way, how self-timed systems, specified by Petri nets at different levels of abstraction, can be expressed in VHDL and simulated. This approach, if combined with that of I. Blunno and L. Lavagno, would provide an idea of a future HDL-based design environment for asynchronous controllers. The front end of this environment, at both specification and feedback sides, is supported by HDL interface, while the back end and appropriate logic synthesis routines are applied to an "object-code"of Petri nets and STGs, suitably hidden from the (possibly inexperienced) designer.

Part V (Architecture Modelling and Performance Analysis) focuses on a generally important application area of Petri nets, the area of performance analysis, with interesting industrial case studies. First, in chapter 13, A. Xie and P. A. Beerel, discuss the performance analysis of asynchronous circuits and systems using Stochastic Timed Petri nets (STPN), where both an overview of the state-of-the-art and original results of the authors in applying STPN to self-timed circuits are presented. They illustrate their method by applying it to Intel's recent asynchronous design experiment, an instruction length decoder for the X86 architecture, called RAPPID. In chapter 14, B.R.T.M. Witlox,

P. van der Wolf, E.H.L. Aarts, and W.M.P. van der Aalst, present a systematic approach to model dataflow architectures at a high level of abstraction using Timed Coloured Petri Nets, with the purpose of performance analysis. They apply the method to the Prophid dataflow architecture, recently studied at Philips. Chapter 15, written by M. Gries, describes a method in which a complex Petri net model of a memory subsystem can be constructed and analysed for performance. The paper illustrates how using Timed Coloured Petri Nets and associated modelling environment (CodeSign) eventually helps to improve a particular memory organisation. Finally, W.M. Zuberek presents a method for the performance modelling of multi-threaded processor architectures with distributed memory. Timed Petri nets are used to model a number of such architectures at the instruction and thread levels.

This book is based on the proceedings of two International Workshops on Hardware Design and Petri Nets, held in 1998 in Lisbon and in 1999 in Williamsburg, VA, as part of the 19th and 20th International Conferences on Applications and Theory of Petri nets (ICATPN).

Acknowledgements. We are very much indebted to all the authors contributing to this book, to Jordi Cortadella and Ganesh Gopalakrishnan, the invited speakers of the Workshops, to our colleagues who helped us with reviewing this material at the pre- and post-Workshop stages, namely Albert Koelmans, Maciej Koutny, Ken McMillan, Marta Pietkiewicz-Koutny, and of course to all the participants of both HWPN workshops for their constructive criticism. We also address our special thanks to the Steering Committee of the ICATPN for offering us the opportunity to organise the Workshops, to Gianfranco Ciardo, the ICATPN'99 Organising Committee Chair, and all other colleagues from Universidade Nova de Lisboa and from the College of William and Mary for the local organisation of the workshops. Many thanks to Kluwer Academic Publishers, and in particular to James Finlay for efficient handling of this book.

ALEX YAKOVLEV

LUIS GOMES

LUCIANO LAVAGNO

I
HARDWARE MODELLING USING PETRI NETS

Chapter 1

COMPREHENSIVE CAUSAL SPECIFICATION OF ASYNCHRONOUS CONTROLLER AND ARBITER BEHAVIOUR

A Generalized Signal Transition Graph

Ralf Wollowski and Jochen Beister

Department of Electrical Engineering, University of Kaiserslautern, P.O. Box 3049, D-67653 Kaiserslautern, Germany, {wollo,beister}@rhrk.uni-kl.de

Abstract: The topic of this paper is the precise modelling of all known forms of asynchronous controller and arbiter behaviour from a causal point of view, and for the purpose of synthesis. In addition to the causal relations between input and output edges provided by the conventional STG (dependence, independence, and exclusion) two forms of pseudo-causality, b-and tcb-concurrency (in signal-tracking behaviour), causal linkage (in multiple input changes and bursts), and race causality (temporal relations as causes, in critical input races) are found to be needed. For their Petri net representation, tc-labelled read and inhibitor arcs, transitions labelled with joint events, and decision transitions representing internal events are introduced. Based on these results, a generalized STG for the first time allows precise causal specifications of all known forms of arbiter behaviour, in particular of "three-way" arbiters that recognize and respond to simultaneous requests with a specific reaction. Several circuit examples are discussed.

Keywords: arbiters, asynchronous circuits, Petri nets, Signal Transition Graphs

1. INTRODUCTION

As event-driven systems, asynchronous circuits react *immediately* and in general *concurrently* to rising and falling edges of their binary input signals x (input edges), namely by generating edges of their output signals y (output edges). An adequate design basis, therefore, would be a specification of the desired behaviour at the XY-interface from a *causal* point of view, i.e. a definition of the dependences between the individual edges instead of their required orderings in time.

3

A. Yakovlev et al.(eds.), Hardware Design and Petri Nets, 3-32.
© 2000 *Kluwer Academic Publishers.*

Focussing on the design of asynchronous controllers and arbiters, the best-known *causal* specification formalism is the Petri net based Signal Transition Graph (STG), e.g. [1,2,3]. Customarily, however, STGs are used only with a specific approach to synthesizing speed-independent (SI) circuits based on Muller´s model [4], and under specific delay assumptions (e.g. [3,5,6]). The STG, therefore, has been confined to reflect certain properties such as commutativity, a property not at all necessary, for example, for extended-burst-mode (XBM) synthesis [7,8,9]. The XBM method, on the other hand, is based on the Huffman model, makes its own delay assumptions, and, especially, uses its own specification formalism: the XBM machine (XBMM) which, being an extended finite-state machine, does not provide a causal viewpoint. So each of the known synthesis methods has its own, not necessarily causal, specification formalism and assures hazard-free implementations by compliance with specific conditions.

This situation is of course unsatisfactory because the designer has to restrict himself *from the beginning* to a *particular* synthesis method with its particular limitations and specification formalism, and may not even have the means to specify the required behaviour from the appropriate, causal viewpoint. Would it not be more natural and convenient to concentrate first (and without a priori constraints) on the *precise causal modelling* of the behaviour *required by the problem* within a *unified* specification formalism, and only *then* to decide which synthesis method is applicable or most appropriate? Non-specialized developers would certainly welcome such a *comprehensive design method* (s. Fig. 1.1). It was already shown in [10] how to derive XBM solutions, if they exist, from STGs, i.e. how to transform an STG into one or - by decomposition - several XBM machines. This suggests that the STG might be a suitable unified causal design entry.

One would expect such a unified model to be *general,* i.e. capable of specifying *every* kind of asynchronous controller behaviour precisely from a causal viewpoint, including indeterminism. However, as shown in section 3, the established version of the STG [3] does not meet this expectation; the causal relations it provides (dependence, independence, exclusion) are insufficient. A complete survey of known asynchronous controller behaviour has revealed additional causal relations: two kinds of pseudo-causality, b-and tcb-concurrency (e.g. required for signal-tracking behaviour), causal linkage (e.g. in multiple input changes and bursts), and race causality (for specifying critical input races in arbiters).

Based on these results, we define a generalized STG (gSTG), and discuss the completeness of the causal relations provided (section 4). We propose this gSTG as the initial causal specification for all kinds of asynchronous controller behaviour, and as the unified design entry of a comprehensive design method.

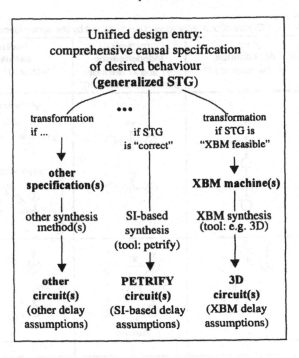

Figure 1.1. Postulated comprehensive design method for asynchronous controllers

2.   BASIC NOTIONS AND CONCEPTS

In this section we discuss the causal relations provided by the currently established STG model and give reasons for using the step graph as the purely temporal model of causally specified behaviour. But first of all we wish to clarify our modelling viewpoint.

2.1   Modelling viewpoint

The models will *specify* required and desired behaviour for the purpose of *synthesizing* - and not primarily analyzing or simulating - a circuit.

We consider systems of *two interacting agents* with *distinct responsibilities*: The *environment* controls the input signals, from which the *controller*, by virtue of its physical structure, generates the output signals. The behaviour of the environment is assumed to be given, sufficiently known in all aspects relevant to the design problem, and not subject to modification. The controller is initially regarded as a black box, with nothing known about internals, especially nothing about internal states and concrete delays.

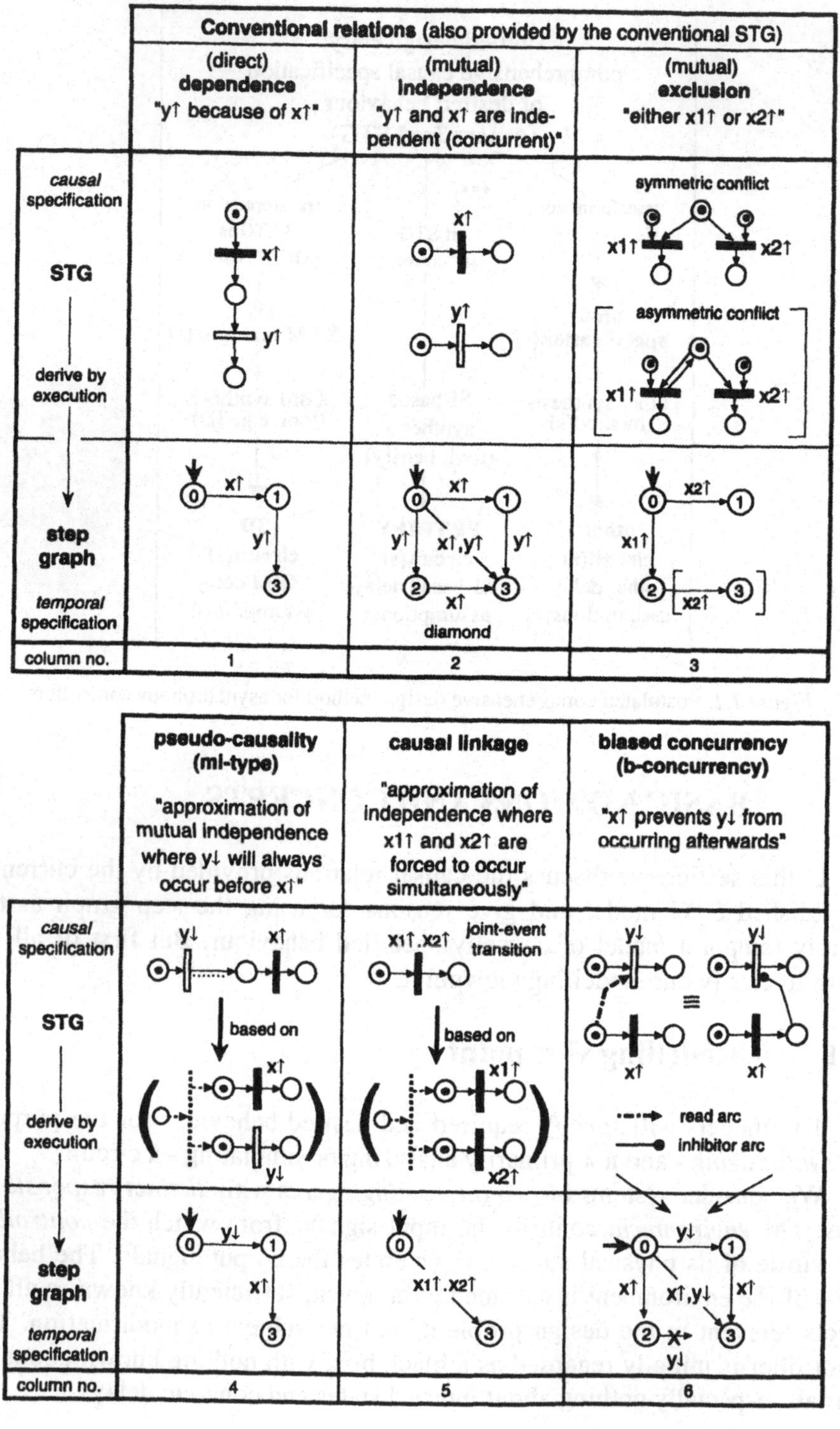

Figure 1.2. Modelling power of the generalized STG model. Part A: Purely causal relations

Asynchronous circuits, being event-driven, react immediately and in general concurrently to their input edges by generating output edges; the latter, in turn, acting through the environment, may cause or at least permit further input edges. Synthesis from the specification will eventually determine the structure of the circuit on the physical (gate or transistor) level, and, based on the laws of physics, this structure will implement causal dependences of output upon input edges. We therefore believe that the adequate design entry should be a *causal* specification of the desired interaction at the input-output (XY-) interface of the controller: a definition of the *dependences* that are to be implemented between the individual input and output edges, and not a specification of their required orderings in time. That is, we prefer the causal viewpoint because we believe it to be the most appropriate for synthesis, and not just because it offers more compact representations.

We require the initial causal specification to be *complete*: Critical behaviour such as *critical input races ought to be represented*; it should not be excluded from the very beginning by *intuitively* partitioning the circuit into an arbiter component dealing with the races and a "noncritical" remainder for which alone a formal specification is then constructed. In short, we propose a *self-contained* specification from which arbiters can be extracted by *systematic decomposition* if necessary. Of course, the implementation of races unavoidably brings up the problem of anomalous behaviour. But these anomalies should not be specified: Their concrete nature is not predictable in general (oscillations, metastability, or spikes, depending on the technology used), and, besides, anomalous behaviour is highly undesirable and has to be "filtered" out. We believe that a good specification of critical behaviour should show precisely what the *desired* behaviour should ideally be like - not least to have a behavioural model against which every implementation can be measured.

Finally, the causal specification should be so *precise* that no additional verbal or textual comments are needed to explain what is really desired.

2.2   Causal relations provided by conventional STGs

STGs are Petri nets where the firing of a transition is interpreted as the occurrence of a signal edge [1,2]. Every transition is labelled with either the rising or falling edge of an input or output signal (e.g. x↑ or x+, y↓ or y-), and will be called an input or an output transition, respectively. All arcs have weight 1, and each place is labelled with a non-negative number of tokens representing its initial marking M_0. A transition t is enabled at a marking M_i whenever each of its preplaces is marked with at least one token. The firing of t is interpreted as an occurrence of the signal edge with which it is labelled. Its currently established form (e.g. [3]) will be called the conventional STG.

Conventional STGs can model causal dependence, causal independence, and exclusion between the input edges (IE) and output edges (OE) of the controller to be designed. Columns 1 to 3 of Fig. 1.2 show one STG example each for these conventional relations. (In contrast to the usual graphical STG notation we do not omit places with one pre- and one posttransition, and use full rectangles for input transitions and hollow ones for output transitions - a style first used in [11]).

A (direct) causal dependence ("edge 2 because of edge 1, E1 → E2") may exist for all combinations of input and output edges: IE1 → IE2, OE → IE, IE → OE, and OE1 → OE2. The environment is responsible for establishing the first two relations, the circuit for the other two. OE1 → OE2 models a multiple output change [12]. Note that direct dependence is physically possible between edges of the same signal only if they have opposite directions.

Causal independence (concurrency) is also possible for all pairings of input and output edges, except edges of the same signal.

The third conventional relation, exclusion, does not offer this freedom of pairing. It models an alternative (free choice or conflict) between several edges. But the choice can only be made, or the conflict solved, if one and the same agent is responsible for each of the conflicting edges and for deciding which edge should actually occur. Exclusion, therefore, exists either only between input edges or only between output edges. Exclusion between an input and an output edge would require a (nonexistent) third "meta-agent" to solve a conflict between circuit and environment.

A useful distinction: if none of the other conflicting edges can occur after the selected edge E, the conflict will be called symmetric. If, on the other hand, one or more of the other edges (again) become possible after E has occurred, the conflict is asymmetric: E excludes the others for the duration of its occurrence only. The asymmetric conflict is modelled either by a two-way arc connection between the E-transition and the place in common to the conflicting transitions (as in Fig. 1.2, column 3) or by labelling several transitions with the identifier of the re-enabled edge (as in Fig. 1.3d, where two transitions are labelled "c↑").

2.3   Step graphs

From the purely causal STG that specifies dependences between signal edges, we can derive a purely *temporal* model specifying all possible orders of the edges in time. Usually, the reachability or state graph is constructed from the STG for this purpose. Every path through this graph (firing sequence) represents such a possible temporal behaviour: linearly ordered sets of single edges in time, consistent with the causal specification. But this graph does not show which edges, being concurrent, might also occur *simultaneously*. (Due to inertia, every circuit has a non-zero time interval - a measure of its resolving power - within which it reliably recognizes and treats input changes as "simultaneous".) This is why we prefer the step graph (SpG): the state graph with additional arcs for the accidentally simultaneous firing of concurrently enabled transitions, i.e. simultaneous signal edges. As shown later, the presence or absence of these additional arcs distinguish certain forms of behaviour from the temporal viewpoint, and therefore they must not be considered as present implicitly, e.g. through implicit transitivity.

Intuitively, a step is any set of concurrently enabled, simultaneously firable transitions; section 4 gives a formal definition. As examples, consider the SpGs shown in Fig. 1.2, derived from the conventional STGs (columns 1 to 3). While the step graph is identical with the reachability graph in case of dependence and exclusion, it differs in case of independence by the additional arc that represents step $\{x\uparrow, y\uparrow\}$ and models the accidental simultaneity of $x\uparrow$ and $y\uparrow$.

3.   THE ADDITIONALLY REQUIRED RELATIONS

The conventional STG can model certain kinds of controller behaviour only imprecisely, others not at all. Therefore, additional causal relations are needed. The present section introduces these new relations, motivates them by examples, and shows how they can be modelled by means of a Petri net, i.e. with a generalized STG model.

3.1   Pseudo-causality of the mi-type

This form of *fictitious* dependence applies between two mutually independent events if it is certain that they will *always* occur in the same *temporal* order. For illustration, suppose that two equally large balls, one of iron and the other of paper, are dropped from the same height at the same time. Both will fall to the earth, neither interferes with or influences the other, but because of the combined action of gravity and air resistance, the iron ball will always hit the ground first.

Consider column 4 of Fig. 1.2, where $x\uparrow$ and $y\downarrow$ are the independent events. Suppose that both are triggered by a common cause, and that y is generated by a fast electronic controller, but x by a comparatively slow mechanical system.

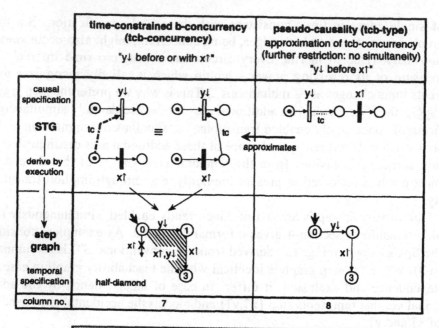

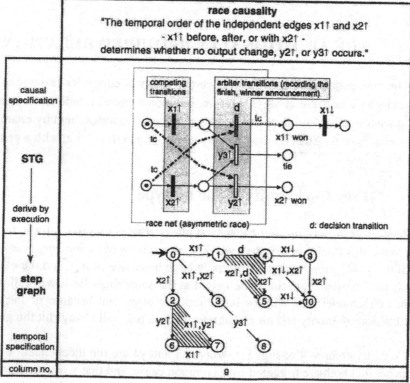

Figure 1.2 (continued). Modelling power of gSTGs. Part B: Temporal-causal relations

Then y↓ will *always* occur before x↑, so that y↓ *seems* to be the *cause* of x↑. It appears appropriate in this case to specify x↑ as causally dependent on y↓ in order to avoid inflation of both specification and circuit, but to indicate the fictitious nature of this dependence by dotting the unlabelled arc from the y↓- to the x↑-transition.

For although x↑ and y↓ are mutually independent in reality (and this is why we call it pseudo-causality of the *mi*-type), there are physical reasons that *cause* them to occur in a fixed instead of random order: the different delays in the controller and its environment. Pseudo-causality may therefore be regarded as a purely causal relation, and hence is listed in part A of Fig. 1.2.

Pseudo-causality of the mi-type is closely related to the concept of difference constraints presented in [6]. However, these constraints are not specified initially but only *subsequently* as the result of the *analysis* of a first (speed-independent) implementation, and between output edges only. Therefore, difference constraints seem to be a special case of pseudo-causality since the latter is applicable from the beginning, also covers the environment, and emphasizes the causal nature of the constraint.

3.2   Causal linkage

Let us repeat the ball dropping experiment, this time on the moon. Again the balls, dropped at the same time, fall without influencing one another. But now there is no air resistance, and the moon´s gravity will *cause* both balls *always* to hit its surface *at the same time*.

Similarly, independent edges of different signals may be *forced* to occur simultaneously (jointly) by equal delays in the circuit. Forcing is causal, therefore forced simultaneity will be called causal linkage. It is feasible only if the edges that are to occur together are triggered by the same signal and generated by components *intentionally* built with approximately equal delays. Therefore, linkage is only reasonable between edges for which the same agent is responsible: either only input or only output edges. Causal linkage of input edges is necessary for implementing multiple input changes in fundamental-mode [12]; linkage between output edges should be specified whenever two or more output signals are required to change at the same time, as in an XBM output burst.

In the generalized STG, causal linkage is modelled by means of a so-called joint-event transition: an input or output transition labelled with the linked edges (separated by a dot). For example, the joint-event transition shown in Fig. 1.2, column 5, specifies causal linkage of the input edges x1↑ and x2↑. In the step graph, the firing of a joint-event transition generates a single arc labelled with the joint-edge identifier, specifying "simultaneous occurrence only". This arc can be seen as the diagonal of the diamond which would result if the cause that forces simultaneity were removed.

Causal linkage is closely related to the concept of simultaneity constraints [6] which is subject to the same restrictions as the already mentioned concept of difference constraints. Therefore, causal linkage is the more comprehensive concept.

3.3   Biased concurrency

Consider the case where an enabled output edge OE will occur unless prevented by an independent input edge IE (non-persistency!). This relation is called biased (or b-) concurrency: concurrency because initially IE and OE are independent, biased because the IE may prevent the OE, but not vice versa. Two features distinguish it from the asymmetric exclusion by which it is usually approximated [13]: Two *different* agents are responsible for the b-concurrent edges, and the OE is unable to exclude the IE for the duration of its own occurrence.

Formally, an input edge IE is biased-concurrent (b-concurrent) to an output edge OE iff

(i)   the environment generates the IE independently of the OE, and

(ii)   the IE *prevents* the OE if the OE has not yet occurred.

That is, the IE is generated without regard for the OE and makes it impossible for the OE to occur afterwards. B-concurrency is modelled by either dot-dash-line read arcs or inhibitor arcs. Column 6 of Fig. 1.2 shows the specification of b-concurrency between the OE $y\downarrow$ and the IE $x\uparrow$ by means of a read arc and an inhibitor arc, respectively. (As formally defined in section 4, tokens do not flow over read and inhibitor arcs.) The figure below shows the associated step graph which specifies the resulting temporal behaviour. Since the IE $(x\uparrow)$, if it is the first to occur, prevents the OE $(y\downarrow)$, no $y\downarrow$-step to marking M3 is possible at M2 (prevention case). On the other hand, due to the independence of the x-edge, $x\uparrow$ can indeed occur after $y\downarrow$ (path M0-M1-M3); $x\uparrow$ and $y\downarrow$ can also occur simultaneously because the IE cannot prevent the OE at the moment of their occurrence (the IE needs time to propagate through the circuit in order to suppress the OE).

We allow both read and inhibitor arcs for purely practical reasons: simpler nets sometimes result if one kind is used instead of the other. The step sequence semantics chosen is called "a-priori concurrent semantics" in [14]. The introduction and use of "read arcs" instead of the established "test arcs" is necessary because the latter exclude simultaneity [15].

B-concurrency is a form of "weak causality" [16] interpreted in the context of asynchronous behaviour. We call it "b-concurrency" instead of "weak causality" because this name is nearer to the essence of this critical relation:

the IE is *unilaterally* independent of the OE. This is crucial to show because it can cause anomalous behaviour if the IE occurs just as the circuit is busy generating the OE. (But these undesired anomalies should not be specified, as discussed in section 2.1.) - To the best of our knowledge, all other concepts that are related to b-concurrency, like asymmetric conflicts ([17], [18]) or not-causality (e.g. [19]), do not capture its essence: unilateral independence.

3.4   Time-constrained b-concurrency

For the specification of required behaviour, b-concurrency is of minor importance since one usually does not wish the OE to be prevented; in general, the designer is urged to ensure that both edges occur. Therefore, we define time-constrained b-concurrency (tcb-concurrency); it exists between an input edge IE and an output edge OE iff
(i)   the IE is b-concurrent to the OE, and
(ii) the additional requirement exists to make it as certain as possible that the OE, although it may be prevented in principle, actually *takes place*.
I.e. the OE has to be generated so quickly that it occurs *before, or at the latest simultaneously with* the IE. In other words: The physically possible prevention of the OE by the IE is declared to be undesirable. Since tcb-concurrency also addresses temporal aspects, it belongs not to the purely causal relations but to the so-called temporal-causal ones (part B of Fig. 1.2). Column 7 shows how to model tcb-concurrency between $x\uparrow$ and $y\downarrow$ in the generalized STG model. The specification only differs from b-concurrency in that read and inhibitor arcs are labelled "tc". The resulting step graph is a triangular half-diamond: the OE ($y\downarrow$) may occur prior to and simultaneously with the IE ($x\uparrow$), but the IE as the first event - which would prevent the OE - is not specified because it is unwanted (therefore: $y\downarrow$ "before or with" $x\uparrow$). Here it becomes obvious why we use step graphs instead of reachability graphs: the latter (obtainable by omitting the $\{x\uparrow,y\downarrow\}$-arc) would suggest that the OE *causes* the IE.

An example: the clock suppression circuit
One application where tcb-concurrency is needed is the modelling of signal-tracking behaviour. For example, such behaviour appears in certain situations at the interface of the clock suppression circuit shown in Fig. 1.3a. This circuit gates clock pulses: The output signal, y, is required to follow input signal c whenever input s is at 1 when $c\uparrow$ occurs, and to remain at 0 otherwise. Input signal s is generated without regard for y, and the clock suppressor is unable to slow down the clock. The timing diagram in Fig. 1.3b shows typical behaviour.

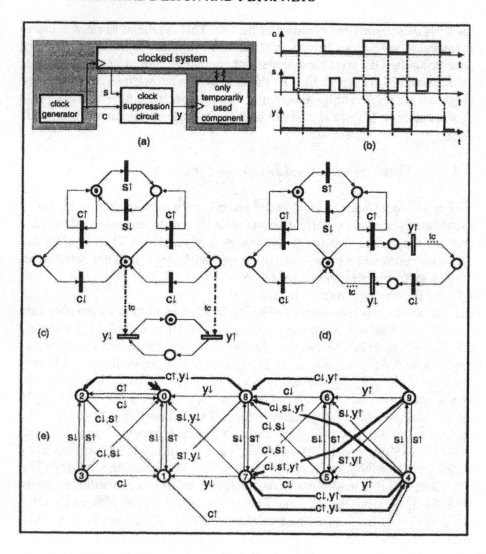

Figure 1.3. Clock suppression circuit with partially restricted input changes

Let us consider the situation where the output signal y is required to follow the input signal c; i.e. where the circuit should wait for a c↓-edge and respond by generating a y↓-edge. The point is that the environment generates c without regard for y; in particular, it will produce the next c-edge (c↑) whether the circuit has produced y↓ or not. What, then, is the relation between c↑ and y↓?

As illustrated in Fig. 1.4, it is

- neither mutual independence (because c↑ prevents y↓ from occurring afterwards, and a level-sensitive circuit can no longer generate y↓ once c↑ has occurred: c = 0 simply did not last long enough),
- nor the causal dependence of c↑ on y↓ (because c is generated without regard for y),
- nor an asymmetric conflict (because c↑ and y↓ are generated independently by different agents, and there is no meta-agent that would decide the conflict and suppress c↑ at least for the duration of y↓).

So the true relation between c↑ and y↓ in the resulting *circuit* is b-concurrency. However, prevention of y↓ would contradict signal tracking and hence is undesirable. Therefore, y↓ and c↑ are *specified* as tcb-concurrent (row 4 in Fig. 1.4).

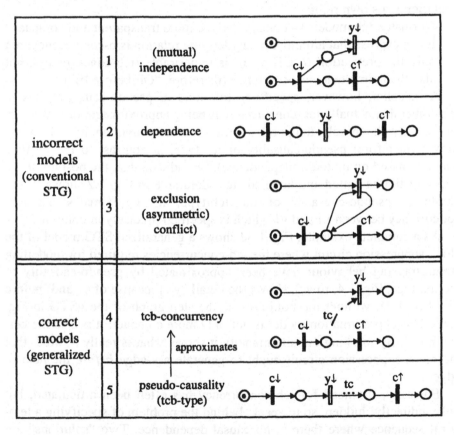

Figure 1.4. Signal-tracking behaviour: incorrect and correct models

Of course, c↓ and y↑, too, are tcb-concurrent. Fig. 1.3.c shows a generalized STG that completely specifies the desired clock-suppressor behaviour. Fig. 1.3e is the step graph (all arcs valid). - Note that this STG model is based on the assumption that the clocked environment of the suppressor ensures that setup and hold times are respected. Hence $c\uparrow$ and $s\uparrow$ (or $s\downarrow$) are not mutually independent and cannot occur simultaneously. The proper relationship between $c\uparrow$ and the s-changes, therefore, is exclusion: the environment generates either $c\uparrow$ or $s\uparrow$ ($s\downarrow$) - see marking M_0 (M_1).

3.5   Pseudo-causality of the tcb-type

This relation can be used for approximating tcb-concurrency by conventional means: An input edge IE is tcb-concurrent to an output edge OE, but we *pretend* that the OE causes the in fact independent IE. There are two good reasons for making this approximation available and to emphazise it as a relation in its own right:

1. Normally, the models so constructed are more transparent and compact.
2. It can express that although the underlying relation is b-concurrency, not only the prevention of OE by IE is undesirable (tcb-concurrency), but also their simultaneous occurrence (therefore: "OE before IE").

A modeller, however, specifying this kind of pseudo-causality, has to remember that actually tcb-concurrency is being approximated, or restricted, and merely the purely temporal aspect - without simultaneity - is being shown (therefore: pseudo-causality of *tcb*-type, in contrast to the mi-type which is based on mutual independence). To indicate this, the postarc of the OE-transition is dotted and labelled "tc". Column 8 in Fig. 1.2 shows an example of pseudo-causality of the tcb-type; it approximates the tcb-concurrency between $x\uparrow$ and $y\downarrow$ which is specified precisely in column 7.

As a complete example, Fig. 1.3d shows a generalized STG model of the clock suppression circuit where the tcb-concurrencies required for modelling signal-tracking behaviour have been approximated by pseudo-causality of the tcb-type; Fig. 1.4, row 5, shows the detail "$y\downarrow$ (because of $c\downarrow$ and) before $c\uparrow$". Fig. 1.3e, without the bold arcs, is the step graph of the gSTG in Fig. 1.3d. This approximation model is not only more compact; it is also the better model. By masking out simultaneity it shows what is really desired: that the clock suppression circuit ought to generate a y-edge *before* the next c-edge.

Basically, the need for tcb-concurrency has often been articulated, but only indirectly; hidden, so to speak, behind the problem of specifying a temporal sequence where there is no causal dependence. Two "solutions" are known for this problem:

1. Inserting a purely temporal relation into the causal scheme (e.g. via "timing constraints" [13]). But then the relation is considered only superficially, as in the case of pseudo-causality, and without showing the

underlying causal concept: the problem has not been addressed from the appropriate, causal side, but "approached" from its temporal aspect.

2. Giving the output transition priority over the input transition (e.g. [20]). This is a merely formal or technical "solution" and not satisfactory because it creates the false impression that the *same* agent is responsible for *both* the IE and the OE, knows their priorities, observes that both are enabled, and generates the output edge first.

3.6   Race causality

Input races in asynchronous circuits are called <u>critical</u> if the required reaction of the circuit depends on the *random order of arrival* of mutually *independent* input edges. Asynchronous circuits with critical races are called <u>arbiters</u>. The conventional STG model is inadequate to specify such race behaviour: certain kinds of races can be modelled only imprecisely, others not at all. This will be shown in 3.6.1 by an introductory example, motivating the definition of a new relation: race causality, defined in 3.6.2; 3.6.3 gives a classification of race behaviour, and 3.6.4 presents more examples.

3.6.1   Introductory example: mutual exclusion

A mutual-exclusion (ME) element controls the exclusive access of two independent users to a shared resource. The required behaviour is the following:
-   users should be granted access in the order of their requests, and
-   simultaneous requests are to be served indeterministically.

Thus, ME behaviour includes critical races between the independent request edges $r1\uparrow$ and $r2\uparrow$. If $r1\uparrow$ occurs before $r2\uparrow$, the ME element is to react immediately with $a1\uparrow$; if, instead, $r2\uparrow$ occurs first, $a2\uparrow$ is required. In case of simultaneity, either $a1\uparrow$ or $a2\uparrow$ has to be chosen at random. Fig. 1.5a depicts the commonly used conventional STG model for ME behaviour, Fig. 1.5b its step graph.

But the STG model does not strictly prescribe the desired dependence upon the order of arrival; it allows the firing sequence $\{r1\uparrow\}$ $\{r2\uparrow\}$ $\{a2\uparrow\}$, i.e. even though the $r1\uparrow$-transition fires first, the $a1\uparrow$-transition does not *necessarily* fire; instead, the $a2\uparrow$-transition may fire. Similar behaviour is exhibited by the permissible firing sequence $\{r2\uparrow\}$ $\{r1\uparrow\}$ $\{a1\uparrow\}$. Furthermore, pending requests should take precedence over later ones, but again this is not what the STG specifies. For example, the step sequence $\{r1\uparrow,r2\uparrow\}$ $\{a1\uparrow\}$ $\{r1\downarrow\}$ $\{a1\downarrow\}$ $\{r1\uparrow\}$ $\{a1\uparrow\}$ is fireable, modelling that even though $r2\uparrow$ occurred at the same time as $r1\uparrow$, the request granted directly after $a1\downarrow$ is not the pending one of user 2, but instead a second r1-request that arrived in the meantime.

The conventional model is fundamentally unable to give a precise specification of ME behaviour since the desired dependence upon the order of arrival cannot be expressed by conventional relations. But what kind of relation do we need? In our opinion, a new causal relationship where the *cause* is a *temporal* relation between events. Consider the case where r1↑ occurs first: a1↑ is not caused by r1↑ *alone* (which is what the conventional model specifies), it occurs because r1↑ has occurred *but r2↑ has not* (if at all, r2↑ will occur after r1↑). Thus, the temporal relation "r1↑ before r2↑" might be regarded as the cause of a1↑. Moreover, using a temporal relation as a cause makes it possible to specify arbiters which explicitly recognize simultaneity of the request signals and respond with a characteristic reaction. - We first define this new relation (race causality) and then show (in 3.6.4) how to use it for constructing precise models of arbiter behaviour, in particular of ME behaviour.

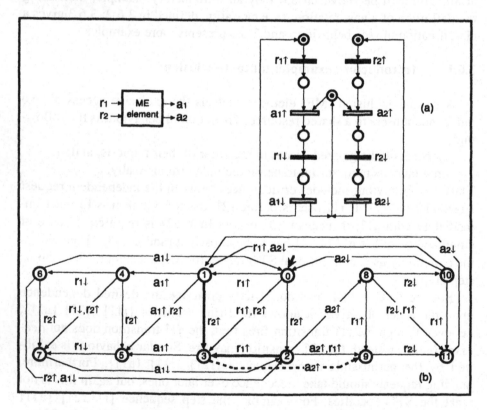

Figure 1.5. Modelling ME behaviour. (a) Imprecise STG model for ME behaviour, (b) step graph for the STGs shown in Fig. 1.5a (all arcs), in Fig. 1.8a (without the bold arcs) and in Fig. 1.8b (without the dashed and bold arcs).

3.6.2   Defining race causality

Let IE1 and IE2 be two mutually independent input edges of an asynchronous circuit to be designed.

Race causality is present iff depending on whether (accidentally) IE1 precedes IE2, IE2 precedes IE1, or IE1 and IE2 occur simultaneously, the circuit is to react with REA1, REA2, or REA3, respectively.

In other words: REA1 occurs *because* IE1 occurred *and* IE2 occurred after IE1 (or not at all), REA2 *because* IE2 occurred *and* IE1 occurred after IE2 (or not at all), REA3 *because* IE1 and IE2 occurred together.

We propose to denote race causality as

(IE1 $\cdots$> IE2) $\rightarrow$ REA1, (IE2 $\cdots$> IE1) $\rightarrow$ REA2, and (IE2 <$\cdots$> IE1) $\rightarrow$ REA3.

Thus, race causality is a causal relationship where the cause is not an event or a set of events, but a *temporal* relation between events. It also is a relationship between more than two events.

Two important questions concerning race causality arise.
1. Can race causality be represented by a Petri net? Yes, by a module or macro (a "race net", see Fig. 1.2, column 9) using only net elements that have already been introduced; no new extensions are necessary. This module will be explained below; it demonstrates that race causality can be seen as composed of already known relations.
2. Is race causality a form of causality in its own right, or merely a form composed of elementary forms? This question is still open. If race causality is indeed an independent form, then it would be justified to introduce a new Petri net symbol for the race-net module.

To demonstrate the development of the module, let us consider a race where the outcome ($x1\uparrow$ before, after, or with $x2\uparrow$) determines whether no output change (REA1), $y2\uparrow$ (REA2), or $y3\uparrow$ (REA3) occurs, and let us try to specify this by a Petri net using the already introduced relations only. This is not trivial because the net has to specify that REA1 (2, or 3) has to occur *if and only if* $x1\uparrow$ occurs before (after, or with) $x2\uparrow$. To solve the problem, we have to determine which physical dependence exists between each pair of edges. For example, in order to react *reliably* with $y2\uparrow$ if and only if $x2\uparrow$ occurs first, an arbiter
1. must start to generate $y2\uparrow$ as soon as it recognizes $x2\uparrow$ (direct causal dependence), and
2. must have generated $y2\uparrow$ before or at the latest with $x1\uparrow$: tcb-concurrency between $x1\uparrow$ and $y2\uparrow$ is called for (otherwise, depending on the state of the arbiter when $x1\uparrow$ occurs, there may be anomalous behav-

iour, or y2↑ may even be suppressed and y3↑ generated instead, i.e. a tie announced).

These considerations lead to the net model in Fig. 1.6, consisting of the elementary relations mutual independence (x1↑,x2↑), dependence (x2↑,y2↑), and tcb-concurrency (x1↑,y2↑). Of course, it is an idealized model. Due to the independence of x1↑ and x2↑ as well as the finite reaction time of a real circuit, the requirement that the prior occurrence of x2↑ *always* should be recognized before, but at the latest at the same time as x1↑, cannot be met in every case; in fact, the true relation in the arbiter is b-concurrency and, therefore, the two edges may be within a critical distance to each other, which inevitably leads to anomalous behaviour. However, with its "before or with" *requirement*, the model stipulates what is ideally required, and therefore constitutes a suitable specification basis against which every implementation can be measured, and points out to the designer the exact locations of the parts most critical in implementation. - In the same manner, of course, it is possible to specify that REA1 is required if and only if x1↑ occurs first. This leads to a net with two "cross-coupled" tcb-concurrencies, which can easily be extended to the complete race net shown in column 9 of Fig. 1.2.

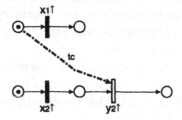

Figure 1.6. Detail of a race net

However, the modelling of REA1 poses a problem because no output change is required; instead, the arbiter is to await the falling edge of x1. Now in any case, and in any implementing circuit, the arbiter must detect and immediately record the outcome by an irreversible change of physical state, and this change must be represented in the net. If an output change (y2↑, y3↑) is to occur as an immediate outward sign of the internal change, then the output change can represent the internal change. But if the outcome is not to be announced immediately, then the position that only signals observable at the interface have to be considered must be abandoned, and the internal decision must be modelled by a new kind of transition, a so-called deci-sion or d-transition, represented by a shaded rectangle labelled "d". In the example, a d-transition must be introduced to remove the token from the postplace of the x1↑-transition; otherwise, the firing sequence {x1↑} {x2↑} {y3↑} would be allowable: the announcement of a tie when actually x1↑ won the race!

The d-transitions should not be confused with dummy transitions; they represent real internal events necessary in every circuit that solves the problem. On the other hand, this internal event may be merely the charge change of a depletion layer. It would therefore be wrong to model it from the beginning as an edge of an internal state signal explicitly introduced for this purpose.

It remains to clarify the relation between the internal decision event d, not observable at the interface, and the *falling* edge of x1. Since a level-sensitive circuit is physically incapable of recognizing x1↑ once x1 has returned to 0, x1↓ would prevent d, and so d ought to occur before or or at the latest with x1↓. Therefore, the correct relationship between the two is tcb-concurrency. In Fig. 1.2 this is approximated by a pseudo-causality of the tcb-type.

From the step graph of the race net (in column 9 of Fig. 1.2) it can be seen that the specified reactions are truly characteristic of the temporal order of x1↑ and x2↑ in the way desired: no output change (or y2↑, or y3↑) is to take place *if and only if* x1↑ precedes x2↑ (or x2↑ precedes x1↑, or x2↑ and x1↑ occur simultaneously). Note that a reachability graph would be unable to represent the transition from M0 to M3.

3.6.3   Classifying race behaviour

In order to be able to describe the known forms of race behaviour concisely but also intuitively, we first create a terminology based on the analogy to a track-and-field race.

The runners are the mutually independent input edges, the so-called <u>competing edges</u>. The occurrence of a competing edge corresponds to a runner's crossing the finish line. The first to occur (to finish) can be either a single edge or a group of simultaneously occurring edges. The arbiter *immediately* identifies the edge or edges that came first (it <u>records the finish</u>) and, also *immediately*, decides how to proceed on the basis of this information alone; i.e. it reacts with REA1, REA2, or REA3 without waiting for the remaining edges to occur and without regard for their placement. Having recorded the finish, the arbiter can proceed in one of two possible ways.

1. It immediately produces a characteristic output change that can be meaningfully interpreted as the <u>proclamation of the winner</u>, and usually is a single edge of a winner-announcing (output) signal (e.g. y2↑ announces the victory of x2↑ in column 9 of Fig. 1.2). In the case of several simultaneous first edges the arbiter does not necessarily have to announce a tie: one may demand that *only one* of these edges be proclaimed the *only* winner. It is, therefore, necessary to distinguish strictly between the recording of the finish and the subsequent <u>determination of the winner</u>.

2. No winner is proclaimed, i.e. there is no (immediate) output change that characterizes the winner; instead, the process is continued with further input changes, and the influence of the outcome of the race on the continuation of the process becomes evident only later, by certain characteristic outputs. In this case we need a d-transition for Petri net representation; cf. the race net in column 9 of Fig. 1.2 if x1↑ wins.

An investigation of all known forms of race behaviour results in the classification shown in Fig. 1.7 (for a detailed discussion see [23]). Section 3.6.4 presents examples of every race type; the classification scheme already contains the names of these examples in brackets. The race described in 3.6.2 (column 9 of Fig. 1.2) is of the asymmetric (deterministic) type.

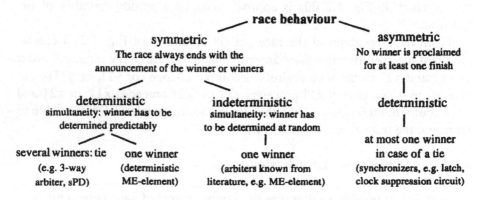

Figure 1.7. Classification of race behaviour

3.6.4   Precise modelling of races: arbiter examples

We demonstrate the capability of the gSTG to model all forms of races precisely by means of a race net specifying the underlying race causality. We also show the deficiencies of the "approximation" by a conventional STG model - provided one is known.

3.6.4.1   Example 1. The ME-element: a symmetric two-way arbiter

Figs. 1.8a and b show generalized STG models that specify ME behaviour more precisely than the one in Fig. 1.5a. Their step graphs are shown in Fig. 1.5b. As required, both models specify that a victory of r1↑ (r2↑) is to be announced by a1↑ (a2↑) only, by means of a race net (set off in grey).

There are differences in the desired reaction to simultaneity. The first model specifies an indeterminate decision as to who is the winner (indeterministic race), and therefore precisely describes the required ME behaviour formulated in 3.6.1, and specifies the currently available ME elements most accurately. The second model, on the other hand, *always* prescribes an a1↑-edge in case of simultaneity (deterministic race).

In particular, both models specify that pending requests should be served as soon as possible. Step sequences such as {r1↑,r2↑} {a1↑} {r1↓} {a1↓} {r1↑} {a1↑} no longer are fireable, but instead sequences such as {r1↑,r2↑} {a1↑} {r1↓} {a1↓} {r1↑,a2↑}.

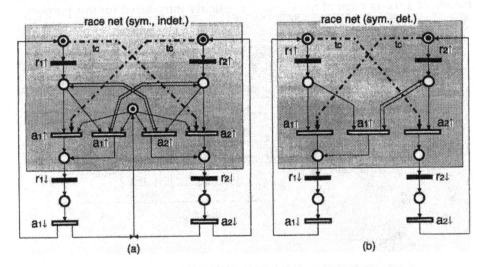

Figure 1.8. gSTG model for (a) interministic (b) deterministic ME behaviour.

3.6.4.2 Example 2. The simple phase detector: a symmetric deterministic three-way arbiter

Systems that process video signals require circuits which create phase balance between independently generated periodic binary signals. One task of such a phase synchronizer is to determine the actual phase displacement and explicitly recognize zero displacement - a task which, due to the mutual independence of the signals, suggests an asynchronous solution. The signals to be balanced would have to be fed to such an asynchronous phase detector, and the detector would have to indicate whether phase balance has already been achieved or which signal precedes or lags behind which other signal; i.e. it would permanently have to arbitrate critical input races. To be in a better position to analyse individual races, we will examine a simplified variant, the simple phase detector (sPD), which determines the phase displacement between two individual 1-pulses in consecutive arbitration cycles. The pulses are generated by two pulse generators PG1 and PG2 (Fig. 1.9a). Independently they produce one pulse each on the input lines R1 and R2 of the sPD. (The points of time at which the two pulses are generated are selected at random.) Now the detector has to determine the phase displacement, i.e.

whether "R1↑ has preceded R2↑" or "R2↑ has preceded R1↑" or whether "R1↑ and R2↑ have (accidentally) occurred simultaneously", and it is required to display the result via (W1,W2,T) = (1,0,0), (0,1,0,), and (0,0,1), respectively. R1↑ and R2↑, therefore, are the competing edges, and W1↑, W2↑, and T↑ ("tie") announce the winner. The races are of the symmetric and deterministic type, and in case of simultaneity a tie is proclaimed by means of a rising edge of the signal T explicitly introduced for this purpose.

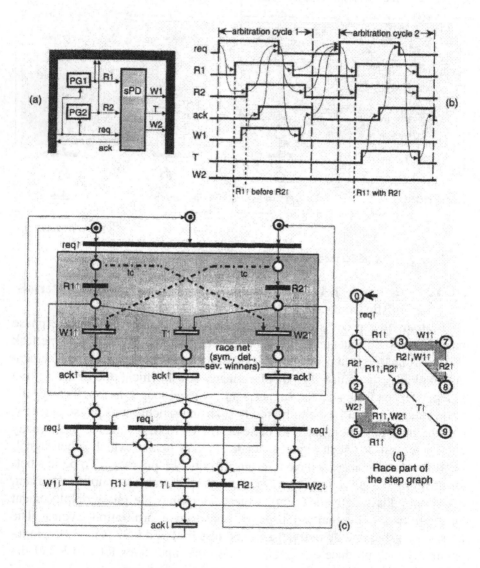

Figure 1.9. The simple phase detector (sPD)

The detection result is to be returned by handshake via signals req and ack. The timing diagram in Fig. 1.9b shows typical temporal behaviour for two consecutive arbitration cycles, where "R1↑ precedes R2↑" during the first cycle, and a "tie" is proclaimed in the second. The detection of simultaneity, therefore, is of paramount importance, as it signifies that the goal of synchronization has been reached; whereas in the case of an individual victory the respective phase shifters would have to be activated.

The adequate representation of these races, i.e. to specify that W1↑ (W2↑, T↑) is to be generated if and only if R1↑ has preceded R2↑ (R2↑ has preceded R1↑, R1↑ with R2↑), is not possible with conventional STGs; they cause the same fundamental modelling problem as with the ME-element, namely to model the "if and only if" requirement.

Fig. 1.9c shows a suitable gSTG model of the desired sPD-behaviour. Set off in grey is the race net, specifying the symmetric and deterministic races between R1↑ and R2↑. Fig. 1.9d shows the part of the step graph reflecting the race (the complete step graph has 34 nodes). Obviously, the "if and only if" requirement is fulfilled.

3.6.4.3   Example 3. Suppressor, flip-flops: asymmetric arbiters

Let us consider another version of the clock suppressor already presented in section 3.4. The difference is that the two input signals are now assumed to be generated independently: the clocked environment of the circuit no longer ensures that setup and hold times are respected. In particular, the suppressor now is required to react to accidentally simultaneous c↑- and s-changes as if s had already assumed its new value ("new s-value counts"). As in the original version, the environment generates c without regard for y; hence signal-tracking behaviour is required. Fig. 1.10a shows a typical run.

Suppressor behaviour includes races between c↑ and both s-edges. The *temporal position* of an s-edge relative to c↑ determines whether c↑ occurs when s = 0, s = 1, or simultaneously with the s-edge, and this in turn determines whether the circuit should react by suppressing or gating the current clock pulse. The races are asymmetric in each case since the output should change only if c↑ occurs. Moreover, since there is no synchronization between the c- and s-changes, the races overlap and merge into a continual race. Fig. 1.10b shows a generalized STG model (STG_CSCnew).

Mainly, STG_CSCnew consists of two race nets (shaded in grey). RN_1 models the races between the c↑ and s↓, RN_2 those between c↑ and s↑. For a more compact representation, both read and inhibitor arcs are used. Transition c↑ belongs to both race nets because races overlap. Because the race is continual, every single c↑- and s-edge as well as every simultaneous change of s and c is recorded and, if necessary, answered by announcing a winner.

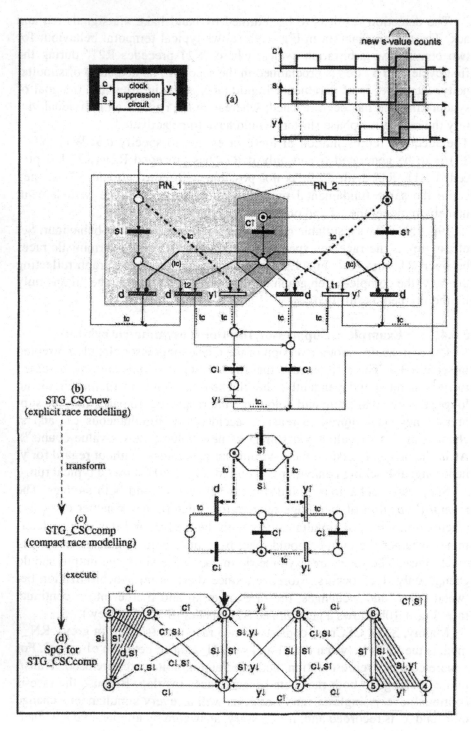

Figure 1.10. Clock suppression circuit with independent input signals.

Recording of simultaneity is modelled by means of transitions t1 and t2. As required, the "new s-value counts", i.e. if c↑ and s↑ (c↑ and s↓) occur simultaneously, the suppressor is required to react as if s had already changed to 1(0): with y↑ in the former case, and no output change in the latter. - Again, signal-tracking behaviour is approximated by pseudo-causality of the tcb-type (cf. Fig. 1.3d).

However, the model is rather complex and not very transparent. Fortunately, it is possible to specify the required suppressor behaviour by the much more compact net STG_CSCcomp shown in Fig. 1.10c; its step graph is shown in Fig. 1.10d. This is due to the reaction desired in case of simultaneity ("new s-value counts"): there is no need to record the victory of an s-edge, since it does not matter whether s↑ or s↓ occurs simultaneously with or before c↑. (For a description of how to transform STG_CSCnew into STG_CSCcomp see [21].)

Compact modelling of asymmetric races between edges of level- and edge-relevant signals is always possible if the *new* value of the level-relevant signal is required to count. In [24] this is demonstrated by two further examples: an edge-triggered latch and a toggle flip-flop. However, the race model of the suppressor cannot be compacted if the *old* value of the level-relevant signal is to count. To specify such behavior in an acceptable, compact way we propose to use a symbol for race causality as a whole (cf. 3.6.2) - but this is future work.

In some few cases it is possible to model asymmetric races not only with a compact race model but also without d-transitions. These are the only asymmetric cases for which one can find conventional STG models in the literature (e.g. [13]); in [24] we demonstrate their deficiencies by the example of a transparent latch.

4.   GENERALIZED SIGNAL TRANSITION GRAPHS

Having described all relations necessary for comprehensive causal modelling together with appropriate means of modelling, we will now merge these results into a formal definition of a generalized STG model.

Definition 1: generalized STG
The tuple gSTG = (**P**,**T**; Firc,C,W,L_T,tc-FR) is a generalized STG if the following holds:
(1) N = (**P**,**T**; Firc) is a net where **P**, **T**, and Firc are non-empty finite sets of places, transitions, and arcs (flow relation), respectively. **T** and Firc are defined as follows.

(1.1) **T** is partitioned into the sets of input, output, and decision transitions.

(1.2) **F**irc is partitioned into
- the set $\mathbf{F} \subseteq (\mathbf{P} \times \mathbf{T}) \cup (\mathbf{T} \times \mathbf{P})$ of ordinary arcs,
- the set $\mathbf{I} \subseteq (\mathbf{P} \times \mathbf{T})$ of ordinary inhibitor arcs,
- the set $\mathbf{I}_{tc} \subseteq (\mathbf{P} \times \mathbf{T})$ of tc-inhibitor arcs,
- the set $\mathbf{R} \subseteq (\mathbf{P} \times \mathbf{T})$ of ordinary read arcs,
- the set $\mathbf{R}_{tc} \subseteq (\mathbf{P} \times \mathbf{T})$ of tc-read arcs,
- the set $\mathbf{C}_{mi} \subseteq (\mathbf{T} \times \mathbf{P})$ of pseudo-causality arcs of the mi-type,
- the set $\mathbf{C}_{tcb} \subseteq (\mathbf{T} \times \mathbf{P})$ of pseudo-causality arcs of the tcb-type.

(2) Function C assigns an unrestricted capacity to each place.

(3) Function W assigns the weight 1 to each arc.

(4) $M_q: \mathbf{P} \to \{0, 1, 2, \ldots n\}$ is a marking of gSTG. M_0 is the given initial marking.

(5) Function L_T assigns a label to each transition. Decision transitions are labelled "d". Input (output) transitions are labelled with either an input (output) edge identifier or a set of linked input (output) edge identifiers.

(6) The tc-firing rule (tc-FR) is applied (as defined in definition 2 below).

Graphically, arcs from **R** are represented by dot-dashed arcs; from **R**$_{tc}$ by dot-dash arcs labelled "tc"; from **I** by the usual inhibitor arc representation; from **I**$_{tc}$ by inhibitor arcs labelled "tc"; from **C**$_{tcb}$ by dotted arcs labelled "tc"; from **C**$_{mi}$ by dotted arcs.

Tc-firing rule

The tc-firing rule has to cope with the "before or with" requirement specified by tcb-concurrency. Before defining it, we give two auxiliary definitions.

- A transition t is said to be <u>activated</u> at a marking M_i if all its preplaces p with $(p,t) \in \mathbf{F}, \mathbf{R}$ or $\mathbf{R}_{tc}$ are marked with at least one token, and all its preplaces p´ with $(p´,t) \in \mathbf{I}$ or $\mathbf{I}_{tc}$ are unmarked.

- A transition $t_r \in \mathbf{T}$ is called a <u>restricting transition</u> of $t \in \mathbf{T}$ if and only if there is a tc-read arc to t_r from at least one preplace p of t with $(p,t) \in \mathbf{F}$, or a tc-inhibitor arc to t_r from at least one postplace p´ of t.

For example, in Fig. 1.10c, the y↑-transition is a restricting transition of the s↓-transition, and the d-transition restricts the s↑-transition.

An *activated* restricting transition t_r imposes restrictions on the firing of t, i.e. it *can* prohibit the firing of t *alone* (without t_r firing simultaneously). This is because the *structural* conditions are such that t and t_r generate tcb-concurrent events. If *no* restricting transition of t is activated at M_i, t is allowed to fire by itself (in Fig. 1.10c: s↓ at M_1, s↑ at M_0); but if restricting transitions of t *are* activated at M_i, then, in general, t can only fire at the same time as (or after) these (in Fig. 1.10c: s↓ with y↑ at M4, s↑ with d at M9). This is expressed by the following definition.

Definition 2: tc-firing rule
Let U be a non-empty set of transitions, Ms a reachable marking, and Ms(p) the number of tokens on place p, and let $|p•(U)|$ be the number of posttransitions of p that belong to U and are connected to p by an ordinary arc ($\in$ **F**). Then U is enabled at Ms and called a <u>step</u> iff
(1) every transition of U is activated at Ms,
(2) for all places p of the net, Ms(p) ≥ $|p•(U)|$ holds, and
(3) firing all the transitions of U does not deactivate any restricting transition tr (of any t) activated at Ms, tr $\notin$ U, unless tr already becomes deactivated by the firing of those transitions ti $\in$ U for which tr is *not* a restricting transition (i.e. unless there is a conflict or a b-concurrency specified between the tr-event and at least one ti-event).
Firing step U means firing all its transitions simultaneously. Firing a transition t leads to a new marking Mi+ with one more token in each postplace of t, and one less token in each preplace of t connected to t by an ordinary arc (p,t) $\in$ **F** (no tokens flow over read and inhibitor arcs).

In comparison, a conventional STG has no decision, joint-event, and restricting transitions, and only ordinary arcs (**F**irc = **F**); condition (3) does not appear in its step definition and firing rule.

Fig. 1.11 shows an example that illustrates condition 3. At the initial marking of the gSTG in Fig. 1.11a, all three transitions are activated. Neither y1↑ nor y2↑ have a restricting transition, but each of them is a restricting transition of x↑. Thus x↑ alone is not allowed to fire first. Further, x↑ is not able to fire simultaneously with *both* of its restricting transitions because they are in conflict; therefore U1 = {y2↑, x↑} and U2 = {y1↑, x↑} are steps, but U3 = {y2↑,y1↑,x↑} is not. Fig. 1.11b shows the resulting step graph.

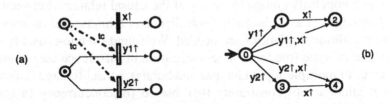

Figure 1.11. tc-firing rule with conflicts between restricting transitions.

Completeness of the provided relations
Of course, having defined a comprehensive modelling formalism, the question of its general validity arises: is the model complete in the sense that

the provided relations are sufficient to model *every* kind of asynchronous behaviour precisely from a causal viewpoint? We are not able to *prove* completeness - and believe that this is a difficult problem since it asks for a sufficient formal definition of asynchronous behaviour. The new causal relations proposed have been found through an empirical study of known asynchronous behaviour. However, we presume completeness, and base this on the following arguments:

1. To the best of our knowledge, *general* investigations of the causal modelling of concurrent systems have not revealed any other relation suitable in the asynchronous context.
2. Critical behaviour - the implementation of which unavoidably brings up the problem of anomalous behaviour - can be precisely specified.
3. OR-type behaviour can be specified (cf. [24]).
4. Diamond-completeness can be shown. Given an SpG diamond expressing mutual independence between two edges, for *each* of its subgraphs consisting only of the source and markings reachable from the source, and including at least one occurrence of each edge, a causal relation exists that has this subgraph for the possible temporal relations. This is shown in [25].

5.   SUMMARY

The topic of this paper has been the *precise* modelling of *all* known forms of asynchronous controller and arbiter behaviour from a *causal* point of view, and for the purpose of synthesis. The work was motivated by the observation that certain kinds of asynchronous behaviour such as critical input races and signal-tracking behaviour cannot be modelled with the causal relations (dependence, independence, and exclusion) provided by the conventional STG. Our aim has been to overcome these inadequacies.

Due to a hopefully complete survey of the causal relations between input and output edges of asynchronous controllers and arbiters, we can now point out which additional relations are needed. We found pseudo-causality of the tcb- and the mi-type (the latter a generalization of difference constraints [6], needed e.g. in multiple input changes and bursts), causal linkage (a generalization of simultaneity constraints [6]), biased (b-)concurrency (a form of "weak causality" [16]), its time-constrained form tcb-concurrency (e.g. in signal-tracking behaviour), and race causality (temporal relations as causes, needed in critical input races).

For Petri net representation of the new relations, several elements beside those of the conventional STG were found to be needed: unlabelled and tc-labelled read and inhibitor arcs, pseudo-causality arcs, transitions labelled with joint events, and decision transitions for modelling necessary but unobservable (internal) reactions of the circuit in case of asymmetric races. - Based on these results, we have defined a generalized STG (gSTG) and have discussed the completeness of the causal relations provided.

This gSTG model has made it possible for the first time to specify precisely all known forms of critical race (or arbiter) behaviour from a causal point of view. In particular, it is now possible to specify the required behaviour of *deterministic* arbiters where a deterministic reaction is desired if several requests accidentally occur simultaneously. This has been demonstrated for the *symmetric* deterministic form by a three-way arbiter (simple phase detector), as well as for the *asymmetric* deterministic form by a suppressor circuit.

Because of its comprehensive causal modelling power, we propose that the gSTG be used as the initial causal specification of the interaction between asynchronous circuits and their environment, and as the unified design entry of a comprehensive design style that enables the designer to choose the appropriate synthesis method and derive his design by appropriate transformations of the gSTG specification (cf. Fig. 1.1). In [24,25] we sketch how to treat the new relations when synthesizing circuits from the gSTG, and show that they do not cause new problems but point out to the designer the exact locations where anomalous behaviour must be expected.

Our work does not aim at more "modelling comfort". This goal - important for the acceptance of Petri net models and for mastering complex designs - has been achieved in [22] for conventional STGs without introducing new causal relations. We propose to combine both forms of generalization.

ACKNOWLEDGEMENTS

The authors thank the reviewers for their truly useful comments and Gernot Eckstein for helpful discussions.

REFERENCES

[1] L.Y. Rosenblum and A.V. Yakovlev. Signal graphs: from self-timed to timed ones. *Proceedings of the International Workshop on Timed Petri Nets*, Torino, Italy, 1985.

[2] T.-A. Chu. On the models for designing VLSI asynchronous digital systems. *Integration: the VLSI journal*, 4, pp. 99-113, 1986.

[3] A. Kondratyev, M. Kishinevsky, and A. Yakovlev. Hazard-Free Implementation of Speed-Independent Circuits. *IEEE Trans. Computer-Aided Design of Integrated Circuits and Systems*, Vol. 17, Nr. 9, pp. 749 - 771, September 1998.

[4]   D.E. Muller and W.C. Bartky. A theory of asynchronous circuits. In *Annals of Computing Laboratory of Harvard University*, 204-243, 1959.

[5]   P. Vanbekbergen, B. Lin, G. Goossens, and H. De Man. A generalized state assignment theory for transformations on signal transition graphs. *Proc. ICCD '92*, MA, Oct. 1992.

[6]   J. Cortadella, M. Kishinevsky, A. Kondratyev, L. Lavagno, A. Taubin, and A. Yakovlev. Lazy transition systems: application to timing optimizations of asynchronous circuits. *Proc. ICCAD '98*, San Jose, CA, USA, Nov. 1998.

[7]   A. Davis, B. Coates, and K. Stevens. The Post Office Experience: Designing a Large Asynchronous Chip. *Proc. 26th Annu. Hawaii Int. Conf. on Systems Sciences*, 1993.

[8]   S. M. Nowick. *Automatic Synthesis of Burst-Mode Asynchronous Controllers*. PhD thesis, Stanford University, 1993.

[9]   K. Y. Yun. *Synthesis of Asynchronous Controllers for Heterogeneous Systems*. PhD thesis, Stanford University, 1994.

[10] J. Beister, G. Eckstein, and R. Wollowski. From STG to Extended-Burst-Mode Machines. *Proc. 5th International Symposium on Advanced Research in Asynchronous Circuits and Systems*, Barcelona, April 1999.

[11] S. Wendt. Using Petri nets in the design process for interacting asynchronous sequential circuits. *Proc. IFAC-Symposium on Discrete Systems*, vol. 2:130-138, Dresden, 1977.

[12] S. H. Unger. *Asynchronous Sequential Switching Circuits*. R. E. Krieger, reprint 1983 (original edition 1969).

[13] J. Cortadella, L. Lavagno, P. Vanbekbergen, and A. Yakovlev. Designing asynchronous circuits from behavioural specifications with internal conflicts. *UPC/DAC TR-RR 94/08*.

[14] G. Chiola, S. Donatelli, and G. Franceschinis. Priorities, Inhibitor Arcs, and Concurrency in P/T nets. *Proc. ICATPN '91*, pp. 182-205, Aarhus, Denmark, 1991.

[15] S. Christensen and N.D. Hansen. Coloured Petri nets extended with place capacities, test arcs and inhibitor arcs", ICATPN '93, *LNCS 691*, pp. 186-205, Springer, 1993.

[16] R. Janicki and M. Koutny. Semantics of Inhibitor Nets. *Information and Computation*, 123, 1995.

[17] G.M. Pinna and A. Poigné. On the nature of events: another perspective in concurrency. *Theoretical Computer Science*, 138, pp. 425-454, 1995.

[18] J.-P. Katoen. Causal behaviours and nets. ICATPN '95, *LNCS 935*, Springer, 1995.

[19] J.-P. Katoen, R. Langerak, D. Latella and E. Brinksma. On specifying real-time systems in a causality-based setting. Formal techniques in real time and fault tolerant systems: 4th International Symposium, *LNCS 1135*, pp. 258-277, Springer, 1996.

[20] R. Janicki. A formal semantics for concurrent systems with a priority relation. *Acta Informatica*, 24, pp. 33-55, 1987.

[21] R. Wollowski. *Entwurfsorientierte Petrinetz-Modellierung des Schnittstellen-Sollverhaltens asynchroner Schaltwerksverbünde*. Doctoral thesis, Univ. of Kaiserslautern, Dep. of Electrical Engineering, 1997 (Shaker Verlag, Aachen, Germany, 1997).

[22] P. Vanbekbergen, C. Ykman-Couvreur, B. Lin and H. de Man. A generalized signal transition graph model for specification of complex interfaces. *Proc. European Design and Test Conference*, 378 - 384, IEEE computer society press, 1994.

[23] R. Wollowski, J. Beister. Precise Petri Net Modelling of Critical Races in Asynchronous Arbiters and Synchronizers. *Proc. HWPN '98* within *19th ICATPN*, Lisbon, June 1998.

[24] R. Wollowski and J. Beister. Comprehensive Causal Specification of Asynchronous Circuit Behaviour: a Generalized STG. *Proc. HWPN '99* within *20th ICATPN*, Williamsburg, June 1999.

[25] R. Wollowski and J. Beister. A Generalized STG. *Technical report B-1-99*, University of Kaiserslautern, Department of Electrical Engineering, 1999.

Chapter 2

COMPLEMENTING ROLE MODELS WITH PETRI NETS IN STUDYING ASYNCHRONOUS DATA COMMUNICATIONS

Fei Xia
Department of Computing Science
University of Newcastle upon Tyne
Claremont Tower, Claremont Road
Newcastle upon Tyne
NE1 7RU, United Kingdom
fei.xia@ncl.ac.uk

Ian Clark
Department of Electronic Engineering
King's College London
The Strand, London
WC2R 2LS, United Kingdom
ian.clark@kcl.ac.uk

Key words: Petri nets, role models, asynchronous communication mechanisms

Abstract: Simpson's role model method [7, 8] was designed for the analysis of synchronisation-free data communication mechanisms employing shared memory and has been shown to be especially useful for the representation and analyses of data freshness properties. Previously published analyses using the role model method have employed proprietary state space search techniques developed by Simpson. Here a formal definition of role models is given and a way of representing role models using Petri nets is presented. Potential advantages of analysing systems using the role model method complemented with Petri net techniques are demonstrated with a case study of data freshness properties of a data communication algorithm.

33

A. Yakovlev et al.(eds.), Hardware Design and Petri Nets, 33-49.
© 2000 Kluwer Academic Publishers.

1. INTRODUCTION

The use of fully asynchronous processes is advantageous in many hard real-time distributed computer systems. For instance, the complete elimination of time interference in data communications between concurrent processes makes it possible to accurately predict the temporal progress of each process in the system because the timing of each one is completely independent. In certain safety critical systems it may also be required that a process cannot be temporally connected to any other processes and must progress at its own pace. In the real time software design method MASCOT, the pool type IDA accommodates the possibility of a complete lack of synchronisation between the reading and writing processes [1, 2].

The class of data communication mechanisms described and studied by Simpson [3] are designed for transferring data between completely synchronisation-free processes. The mechanisms employ between one and four data areas (called *slots*), implied to be in memory shared between the writer and the reader. Some of the mechanisms have logic in the form of control variables to steer the writer and reader to prevent them from accessing the same slot simultaneously and to cause the reader to obtain the most up-to-date data provided by the writer. The reader and writer are not subject to any relative timing constraints and thus are completely temporally independent without the need for synchronisation, wait states or arbitration in the conventional, request/grant, sense. The general organisation of these types of mechanisms is shown in Figure 2.1. They will be referred to as *slot mechanisms* in this chapter.

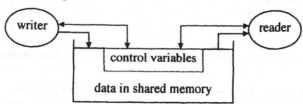

Figure 2.1. Scheme of slot mechanisms.

It is vital, especially in view of the lack of synchronisation, for the slot mechanisms to be proved to pass on data that is both coherent and fresh. Although other solutions to the problem exist [4, 5], Simpson was the first to show a solution using four slots arranged in two pairs to maintain data coherence and data freshness, having found that up to three slots arranged linearly cannot maintain data coherence [3, 6]. In order to prove the data coherence and data freshness properties of the slot mechanisms, Simpson introduced the role model method [7, 8], which is a novel notation describing discrete state transitions to assist a process of exhaustive reasoning and state space search. A proprietary technique, employing what are called *transition diagrams*, which is similar to reachability analysis in Petri nets [9], was proposed by Simpson for the analyses of role models [7].

As a result of the novel nature of the role model method there have been no supporting studies using it for analysis published by other researchers. To date, there has been no attempt to formally define role models. In addition, although it has been claimed that automated, computerised analyses have been carried out based on transition diagrams [8], no software packages by other sources exist by which any analysis result using Simpson's software may be verified.

It has been shown that Petri net models of the slot mechanisms can be obtained and that the loss of data coherence, as it is signified by a relatively simple state (that of the reader and writer accessing the same slot at the same time), can be checked for with the help of the Petri net models in a straight forward manner [10, 11]. However, as data freshness is affected not only by simple states but also by the trajectories of states, the role model method is more useful for its analysis [10, 12, 13].

In this chapter, the role model method is defined in a formal manner. A method of deriving Petri net representations for role models is then given. The intention is to replace or supplement the transition diagram analysis technique with the analysis techniques available for Petri nets to realise more automated and verifiable analyses of asynchronous communication systems. An example is provided where the data freshness properties of a four-slot mechanism algorithm is investigated.

This chapter should be read in conjunction with [3], [7] and [8].

2. ROLE MODELS

The role model method was "developed specifically to deal with the [slot mechanisms] in which small shared control variables are used to co-ordinate access to shared data" [8].

Definition 1. **Role model basics**

A role model is a tuple, $M = (A, E, R, S, s_0)$.

$A = \{a_1, a_2, ..., a_n\}$ is a finite set of *agents* (implied but not formally described in [7]), $n \geq 0$. $R = \{r_1, r_2, ..., r_m\}$ is a finite set of *roles*, $m \geq 0$. $E = \{e_1, e_2, ..., e_l\}$ is a finite set of *events*, $l \geq 0$. The sets of agents, roles and events are disjoint with one another, $A \mid R = \varnothing$, $A \mid E = \varnothing$, $E \mid R = \varnothing$.

An agent may *assume* a number of roles (known as the *role pattern* of the agent [7]). An event may or may not be permitted to *occur*. A *state*, $s_i \in S$, is a complete description of the current role pattern of each agent, in the form of a set of role patterns. Such sets of patterns are generally known as *pattern expressions* or in short *expressions*. In addition, a state also includes an implied specification of which events may occur next.

An event's occurrence may *modify* a state to produce another state. An event e_j, $j \in [1, l]$, is therefore a *state transition function*, a mapping from states to states, $e_j \in E: S \times S$, and is the only provision by which states may be changed.

Events occur according to interleaving semantics, in other words, at most one event occurs at any time, because "events are regarded as distinct" [8].

The dynamic progression of the model consists of the occurrences of events from the *initial state* $s_0 \in S$.

Definition 2. Role patterns

A role pattern, $p_R \subseteq 2^R$, is a subset or a group of subsets of R. A simple way to express it is by writing the names of the *member roles* together in a continuous string. For instance, $r = r_j r_k$ is a role pattern, if r_j and r_k are role names and $r_j \neq r_k$. Agent a_i assuming the roles of r_j and r_k may be written as $a_i r_j r_k$ or $a_i r$ if $r = r_j r_k$.

In addition to role names, the *pattern operators* "*", "0" and "−" may appear in role patterns.

Definition 3. Expressions

A pattern expression (or expression in short) is a set of patterns. Expressions are written in the conventional list style, i.e. member patterns separated by commas.

There is at most one role pattern for each agent of the model in any expression. The simplest non-empty expression is a single pattern consisting of one member role or one pattern operator.

An expression and/or some of its member patterns may be enclosed in (), [], or { }. In general, square brackets indicate that the expression describes a subset of all possible patterns and curly braces indicate that the expression describes all possible patterns within its scope. These and parentheses are used to discriminate between parts of expressions.

The following are example expressions:
$[a_1 0, (a_2 r_4 r_5, a_3 + r_4, a_4*), a_5 + r_1, + e_3 e_1 e_2]$
$[a_1 r_4 r_5*, \{a_2*, a_3*, a_4*\}, a_5 0, e_1 e_3]$
$\{*\}$
where a_i, $i \in [1,5]$, are agents, r_j, $j \in \{1,4,5\}$, are roles, and e_k, $k \in [1,3]$, are events. The parentheses and braces indicate that agents a_2, a_3 and a_4 may represent something distinctive, or are being highlighted for some reason.

Definition 4. Transition statements
A transition statement consists of an *input expression* followed by an →
followed by an *output expression*. It describes an *action*, which, when
carried out, results in a *modification* of the patterns of the input expression
as specified by the output expression.

A role member in the role pattern for an agent in an input expression
indicates that the agent assumes this role before the transition statement. An
"*" in an input expression indicates irrelevance for the transition statement
(don't care). A "0" in an input expression indicates that the agent assumes no
roles.

A role name in the role pattern for an agent in an output expression
indicates the modification of "adding if previously not assumed", i.e. the
specified role is added to the role pattern of the agent if it is not there
previously. If the agent already assumes this role no change is implemented.

An "*" in an output expression indicates no change.

A "‾" on top of a role name in an output expression indicates the
modification of "removing if present", i.e. if the agent previously assumed
the role after the statement it will not assume it any more.

Therefore the basic transition statements, which specify the simplest
modifications, are as follows:

$$[ar''] \rightarrow [ar'] \tag{1}$$

$$[ar''] \rightarrow [a\bar{r}] \tag{2}$$

$$[ar''] \rightarrow [a*] \tag{3}$$

$$[ar''] \rightarrow [a0] \tag{4}$$

The transition statement (1) describes an action of adding the role r' to
the role pattern of agent a, without modifying anything else. After this the
agent assumes the role r' while retaining any other role it had. The transition
statement (2) describes an action of removing the role r from the role pattern
of agent a, without modifying anything else. After this the agent a does not
assume the role r, while retaining any other role it had. The transition
statement (3) describes an action of no change to the role pattern of the agent
a. After this the agent a retains all the roles it had. The transition statement
(4) describes an action of removing all roles from the role pattern of the
agent a. After this the agent a assumes no roles.

Definition 5. Interleaving semantics of transition statements
A transition statement must be constructed in such a way that its overall
modification is obtained by carrying out each of the basic transition
statements it contains according to interleaving semantics (one at a time) in

arbitrary order. In practice, this requires that within any transition statement the same role is not specified to be modified in conflicting ways and effects that depend on the order of actions must be specified using multiple transition statements.

Definition 6. Pattern compatibility

Two patterns p_x and p_y are said to be compatible with each other iff no role is specified present or absent in conflicting ways in p_x and p_y. This relation is denoted by the symbol '~'. The complement relation, incompatibility, is denoted by '/~'.

For instance, $a* \sim ar$ for any r.

Definition 7. Event functions

An event e_i is a state transition function, $e_i \in E: S \times S$ which describes the state change when e_i occurs. It consists of p transition statements, p being a finite number, $p \geq 1$. The description of any event e_i is written as

$e_i: I_{ei1} \rightarrow O_{ei1}, I_{ei2} \rightarrow O_{ei2}, ..., I_{eip} \rightarrow O_{eip}$

where I_{eij} is the jth input and O_{eij} is the jth output expression of e_i, $j \in [1, p]$.

The effect of an event's occurrence is obtained by carrying out the transition statements of the event function one at a time in the sequence specified.

Definition 8. The organisation of events in systems

Events are regarded as atomic, i.e. no event may start when another is occurring. Events may be organised into sequential *processes* in the fashion of conventional programming languages, i.e. by writing them one after another in a sequence. Such sequential processes may also be specified to be concurrent to one another by writing them separately. When more than one event belonging to different concurrent processes are ready to occur at any state, which one occurs next is not deterministic.

3. PETRI NET REPRESENTATION OF ROLE MODELS

Since a role model represents a discrete system with a bounded number of potential states (A, R and E being bounded for any known model) whose actions follow interleaving semantics, a subset of what classical Petri nets can represent [Peterson 1981], it should be possible to find an equivalent Petri net for any role model. Here techniques are developed whereby a safe (1-safe) Petri net representation of any given role model may be found.

Proposition 1. Role patterns
It is proposed to represent the notion of an agent and its possible role patterns with 1-safe places. An agent assuming a particular role in a pattern is represented by a particular place having the marking 1. An agent not assuming a particular role in a pattern is represented by its corresponding place having the marking 0. Complementary places are employed to avoid the need for inhibitor arcs. This necessitates a potential $2m$ such places for each agent in the model and a maximum of $2m \times n$ such places for all agents.

The role pattern of $a_1r_4r_5$, assuming that $m=5$, is thus represented by the Petri net fragment and marking in *Figure 2.2*. The role pattern of a_10 is represented by marking all the (not) places and unmarking all the other places in *Figure 2.2*.

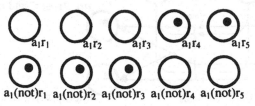

Figure 2.2. Petri net representation of $a_1r_4r_5$.

Rules are provided below for the representation of statements (1) ~ (4). Markings in the figures of this section generally represent the terminal condition of the transition statements (i.e. after the modifications have been carried out).

Proposition 2. Petri net representation of basic transition statements
The transition statement $[ar''] \rightarrow [ar']$ is represented by the Petri net fragment in Figure 2.3. The transition statements (2) ~ (4) are represented by the Petri net fragments in Figure 2.4 ~
Figure 2.6.

There are two types of transitions in Figure 2.3. One type represents the proper modification according to the basic transition statement and include the transitions t_{mod1} and t_{mod2}. The roles of a other than r' need to be referenced by these transitions, depending on if and in what way they are compatible with the pattern r''. If a role is a member of r'', the place representing it must be marked for t_{mod1} or t_{mod2} to fire. If a role is not compatible with r'' the place representing its complement must be marked for t_{mod1} or t_{mod2} to fire. If a role is compatible with but not a member of r'', i.e. it is compatible with r'' solely because r'' contains an *, places representing it are not connected with t_{mod1} and t_{mod2} in any way. These connections represent the fact that the input expression of the transition statement must be satisfied (i.e. it is compatible with the present state) for its action to be carried out. The connections between transitions t_{mod1} and t_{mod2} and the place representing ar' and its complementary place represent the modification of this transition statement. The markings of these places are either modified or retained depending on whether the role r' is assumed by the agent a or not before the action.

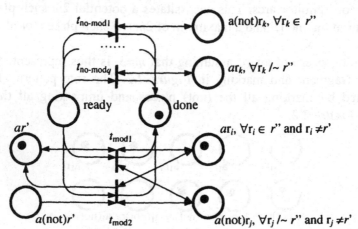

Figure 2.3. Petri net representation of (1).

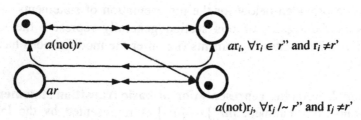

Figure 2.4. Petri net representation of (2).

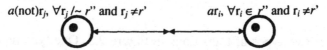

Figure 2.5. Petri net representation of (3).

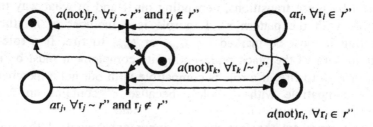

Figure 2.6. Petri net representation of (4).

The "ready" and "done" places are provided so that the sequential nature of transition statements within an event may be represented (below). These places are not explicitly drawn in the other Petri net fragments (Figure 2.4 ~ Figure 2.6) but are assumed to be present. One of the transitions firing in this net fragment takes a token away from place ready and deposits it in place done, and the net fragment only becomes active when ready is marked.

The number of transitions named $t_{no\text{-}modx}$, $x \in [1, q]$ where q is a finite integer are needed to ensure that the net fragment does not introduce deadlocks artificially, i.e. if place ready is marked, at least one transition between it and place done is enabled. At least one of these "no modification" transitions is enabled if a does not assume the pattern r". These transitions are also assumed to be present in all the other net fragments in this proposition but are schematically shown only once in Figure 2.3.

Proposition 3. Representing transition statements

It is proposed that a transition statement t containing a finite number, h, of basic actions be represented with the Petri net model in Figure 2.7.

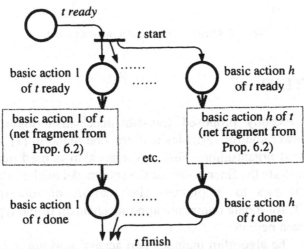

Figure 2.7. Petri net representation of a transition statement.

The structure of Figure 2.7 conforms with the specification that within a transition statement, the basic actions are carried out according to interleaving semantics in arbitrary order.

Proposition 4. Representing events

It is proposed that an event e with a finite number, i, of transition statements be represented by the Petri net model of Figure 2.8.

The place enable in Figure 2.8 is provided so that events are atomic in the model as in the definition. This is a place shared globally among all model nets of events. The start transition of each event model has it as an input place, while it is an output place of the finish transition of each event model. This ensures that when transitions in an event model are firing, no other event model may become active until the current event's finish transition has fired. The optional event ready and done places facilitate the representation of sequential processes in the way described in [9].

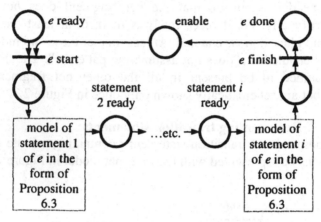

Figure 2.8 Petri net representation of an event.

4. CASE STUDY

The most recently proposed four-slot mechanism algorithm in [8] employs one fewer control variables than the one found in [7]. It also has a more streamlined organisation. This new algorithm is used here in a case study to demonstrate the finer points of the role model method and the use of Petri net techniques to supplement the analysis of role models. The algorithm itself, for a data communication mechanism of the type of Figure 2.1, is reproduced here in

Figure 2.9. The algorithm includes two access procedures, one each for the writer and the reader. Each access procedure is assumed to be embedded within a larger cycle loop in which data to be communicated is produced or consumed. Depending on the arrangement of the larger writer and reader cycles, the access may be open or limited. Taking the writer process as an example, the writer access procedure is assumed to be part of a cycle loop of "loop, produce data, write access procedure, end loop" if it has limited access. If it has open access, the statement *wr* in the access procedure would include the actions of preparation/production of data, i.e. the data is written as it is produced. In this algorithm, the reader and writer procedures are regarded as sequential within themselves, but arbitrary interleaving is allowed between the two processes. Thus it provides for complete temporal independence of the two sides and has been described as "fully asynchronous" or "synchronisation free".

Writer	Reader
wr: $d[ip, w[ip]] := input$	$r0$: $r := w$
$w0$: $ip := \overline{ip}$	$r1$: $op := \overline{ip}$
$w1$: $w[ip] := \overline{r[ip]}$	rd: $output := d[op, r[op]]$

Figure 2.9. Four-slot data communication mechanism algorithm.

Simpson applied role model based analysis on the algorithm in Figure 2.9 and verified its data coherence, data freshness and data sequencing properties [8]. Data coherence issues of the four-slot mechanism have been extensively investigated using Petri net based techniques [10]. In this case study, the issue of data freshness is the focus. Definitions concerning data freshness, obtained from [7] and [8], are given below.

Definition 9. Latest and previous latest data items

At any time, the latest item of data is found in the slot accessed by the last completed *wr* statement, and the previous latest item of data is found in the slot accessed by the last but one completed *wr* statement.

Data freshness issues are relevant only at the beginning of a reader access to a slot, i.e. at the beginning of the statement *rd* in Figure 2.9. Checking for data freshness obviously has to be done at this point.

Definition 10. Data freshness

Data freshness is maintained if the reader always accesses an acceptable slot. Acceptable slots are:

1. the slot containing the item of data which was the latest at the beginning of the pre-sequence of the current reader cycle, $slot_L$;
2. the slot containing the item of data which is the previous latest at the beginning of the pre-sequence of the current reader cycle, $slot_P$, only if the pre-sequence of this reader cycle started while a writer post sequence was in progress;
3. the slot containing the latest item of data before the beginning of the current reader access, $slot_{LL}$.

A reader cycle consists of one cycle of consecutive *r0*, *r1* and *rd* statements. The statements *r0* and *r1* constitute the pre-sequence of the reader cycle they belong to. The post sequence of a writer cycle consists of the statements $w0$ and $w1$.

In Definition 10, 1. is intuitive, as the reader access procedure uses the pre-sequence to determine where the latest item of available data is ($slot_L$), and a successful attempt at obtaining this item of data, even in the event of newer data becoming available during the reader pre-sequence ($slot_{LL} \neq slot_L$), cannot be regarded as a data freshness failure. In comparison, 3. is meant to cover the case in which the reader is after all able to obtain the location of $slot_{LL}$, even when $slot_{LL} \neq slot_L$, by design or by accident. It is recognised in 2. that if a writer post sequence and the pre-sequence of the current reader cycle overlap in time, the reader should not be expected to obtain the location of $slot_L$ because this information may not have been completely indicated by the writer post sequence yet. However the reader should be expected to at least obtain the location of $slot_P$ in this case, as that should have been indicated by the post sequence of the previous writer cycle.

From [10], it is obvious that such information as which slot contains the latest item of data is not directly available from the Petri net models used to study data coherence of the slots mechanisms. These models have only reading, not reading, writing and not writing as the slot related states. Although a careful study of all trajectories generated by a full reachability search should reveal information concerning data freshness properties, extracting it is not a convenient and straight forward process.

Table 2.1. Data freshness related roles.

Roles	Representations
W	slot is being written to
R	slot is being read from
F	slot contains latest item of data known to the writer ($slot_L$ or $slot_{LL}$)
P	slot contains previous latest item of data known to the writer ($slot_P$)
V	slot is $slot_L$ and/or $slot_{LL}$, or $slot_P$ when the pre sequence of the current reader cycle overlaps with a writer post sequence.

More state variables are added in [7] to facilitate the study of data freshness properties using the role model method in the form of data freshness related roles. These make it possible to keep track of $slot_L$, $slot_{LL}$, and $slot_P$ during a state space analysis.

The methodology used in this investigation of data freshness properties of the four-slot mechanism is as follows:

1. Propose a formal definition of role models (not available from [7] but needed for the following step).

2. Develop general techniques whereby a Petri net representation may be found for any role model in an automatic way.
3. Find the Petri net representations of the data freshness related role models in [7] and carry out reachability searches using different software packages.

The roles introduced in [7] for analysing the data freshness properties of the four-slot mechanism are listed in *Table 2.1*.

The Petri net models used in [10] do not have explicit representation of these roles except for *R* and *W*.

The data freshness related events and their transition statements are listed in Figure 2.10.

Writer:

$$wae: \quad [F^*, *, *, *] \to [P, *, *, *]$$

$$[W^*, *, *, *] \to [VF, \overline{F}, \overline{F}, \overline{F}]$$

$$wab: \quad [*, *, *, *] \to [\overline{P}, \overline{P}, \overline{P}, \overline{P}]$$

Reader:

$$rae: \quad [F^*, *, *, *] \to [V, \overline{V}, \overline{V}, \overline{V}]$$

$$[P^*, *, *, *] \to [V, *, *, *]$$

Figure 2.10. Data freshness related events.

The agents in these transitions statements are the slots. The event *wae* occurs at the end of a writer slot access (end of statement *wr*, see *Figure 2.13* below). The event *wab* occurs at the beginning of a writer slot access (beginning of *wr*). The event *rae* occurs at the end of a reader slot access (end of *rd*). During these events the freshness related roles *F*, *P*, and *V* are updated. They should be checked to see if the reader is accessing an acceptable slot during every *rd*.

In Figure 2.10, the absolute positions of the slot agents are flexible, providing for a shorthand of presentation. If *x*, *y*, *z*, and *u* are role patterns, the expression [*x*, *y*, *z*, *u*] represents the situation of one of the slots assuming the *x* role pattern another one the *y* pattern, still another one the *z* pattern, and the last one the *u* pattern. It is not, in this case, indicated which of the slots 00 ~ 11 assumes which of the patterns. This type of shorthand presentation of a role model can only be used if the absolute information about each agent's role pattern is not of significance for the transition statement.

The relative positions of the role patterns are significant, however, in that they are consistent within one transition statement. In other words, once the input expression is written the output expression must correspond to it. For instance, the transition statement

$$[F*, *, *, *] \rightarrow [V, \overline{V}, \overline{V}, \overline{V}] \qquad (5)$$

Represent the action of adding the V role to the slot currently holding the F role and removing the V role from all other slots.

If a reader access statement rd is directed to a slot with the V role data freshness is regarded as being maintained.

Considering Definition 10, 1. is checked for with the V role which was updated at the end of the last reader access (the beginning of the pre-sequence of the current reader cycle) to $slot_L$. Since the writer events do not remove the V role from any slot and this V role is not removed by the other transition statement in the event rae, this remains set until the current rd. For 2. in Definition 10, the P role is set at the end of a writer access but removed at the beginning of the next writer access. In other words, the P role only exists on a slot during the writer post sequence. If the reader pre-sequence starts during such a period, the rae event duly assigns the V role to this slot, signifying the fact that the reader pre-sequence overlaps a writer post sequence hence the reading from $slot_P$ is permitted. Finally, 3. in Definition 10 is managed by giving $slot_{LL}$ the V role together with the F role at the wae event. This ensures that even if a writer access happened immediately before a reader access and after the reader pre-sequence, the reading from the slot just accessed by the writer would be regarded as acceptable.

Petri net models of the events in Figure 2.10 have been developed according to Proposition 1 ~ Proposition 4. Part of the model of transition statement (5) is shown below in Figure 2.11. The no-modification transitions (Figure 2.3) are not shown in Figure 2.11 but are present in the moc els used in the study.

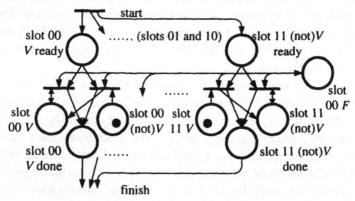

Figure 2.11. Part of Petri net representation of (5).

The Petri net model for the complete four-slot mechanism with additional provision for studying data freshness properties are constructed in the form of Figure 2.12. The monitoring net consists of models of the data freshness related events in Figure 2.10 while the original model used for data coherence analysis in [10] has been retained as the main net. The atomicity of the events in the monitoring net and the integrity of the main net are maintained with a global enable place. Whenever a transition firing in the main net signifies that an event in the monitoring net should start, the token in the enable place is passed to the appropriate part of the monitoring net while both the main net and the other parts of the monitoring net shut down. Since the monitoring net only updates the freshness related role states the main net is not affected by its operations. The monitoring net goes into action at the points within the main net shown in Figure 2.13. Since both *wr* and *rd* statements are regarded as non-atomic in the main net model [10], these are indeed transitions in the main net. The interface between the main and monitoring nets is schematically shown in the example of Figure 2.14.

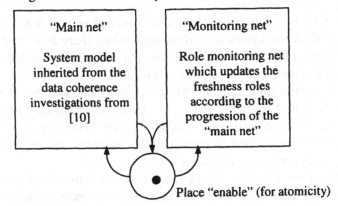

Figure 2.12. Structure of Petri net model used in data freshness study.

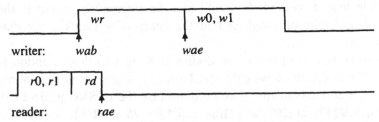

Figure 2.13 Correspondence of main net statements with monitoring net events.

Reachability searches using two software packages developed independently have been run on the data freshness study model. The main net of this model is the version discussed earlier in this chapter based on the non-atomic assumptions for the control variable statements. The results confirm that no reachable marking signifies the reading of a $\bar{V}$ slot. This indicates that the four-slot mechanism maintains data freshness under normal operating conditions, and so supports the results reported in [7].

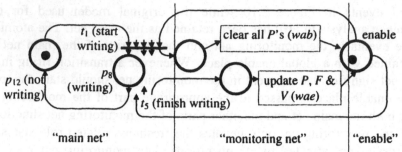

Figure 2.14 Updating data freshness roles on slot $d[0,0]$ for the first writing step *wr*.

5. CONCLUSIONS AND FUTURE WORK

Simpson's role models are presented in a formal way. It has been shown that role models have significant advantages when there is a need to track state transitions over unknown numbers of steps, for instance, when properties such as data freshness are studied.

A way of representing role models using Petri nets has been presented and a general method for using Petri net techniques to complement the role model method has been tried and proved to be viable. Certain finer points of using the role model method in checking slot mechanisms, especially their data freshness properties are explored with a case study.

From related work [14], it seems natural that coloured Petri nets (CPNs) [15] may be used to represent role models in a more straight forward manner, especially considering the shorthand notations of the latter. Therefore this may be a profitable technique to use in future investigations of the role model method. It would also be interesting to see if the role model method may be employed in the study of systems other than slot mechanisms.

This work is part of on-going studies at King's College London (KCL) and the University of Newcastle (NCL) into asynchronous communication algorithms and mechanisms. It is supported by the EPSRC project Comfort (Grant GR/L92471 at KCL and Grant GR/L93775 at NCL).

6. REFERENCES

[1] Joint IECCA and MUF Committee on Mascot (JIMCOM), The official handbook of Mascot, Computing Division, N building, Royal Signals and Radar Establishment, St Andrews Road, Malvern, Worcestershire, WR14 3PS, UK, June 1987. (Now available

from the Defence Research Information Centre, DERA, Kentigern House, 65 Brown Street, Glasgow. G2 8EX.)

[2]   Simpson, H.R., "The MASCOT method", Software Eng. Journal, Vol. 1, (3), pp.103-120, 1986.

[3]   Simpson, H.R., "Four-slot fully asynchronous communication mechanism", IEE Proceedings, Vol. 137, Pt. E, No. 1, pp.17-30, January 1990.

[4]   Kirousis, L.M., "Atomic multireader register", Proc. 2nd Int. Workshop on Distributed Computing, Amsterdam, LNCS-312, pp.278-296, Springer Verlag, 1987.

[5]   Tromp, J., "How to construct an atomic variable, Proc. 3rd Int. Workshop on Distributed Algorithms, Nice, LNCS, Springer Verlag, pp.292-302, 1989.

[6]   Simpson, H.R., "Fully-asynchronous communication", IEE Colloquium on MASCOT in real-time systems, pp.2/1-2/6, May 12, 1987.

[7]   Simpson, H.R., "Correctness analysis of class of asynchronous communication mechanisms", IEE Proceedings, Vol. 139, Pt. E, No. 1, pp.35-49, January 1992.

[8]   Simpson, H.R., "New algorithms for asynchronous communication" and "Role model analysis of an asynchronous communication mechanism", IEE Proceedings on Computing and Digital Techniques, Vol. 144, No. 4, pp.227-240, July 1997.

[9]   Peterson, J.L., Petri net theory and the modeling of systems, Prentice-Hall, 1981.

[10]   Xia, F, Clark, I.G. and Davies, A.C., "Petri-net based investigation of synchronisation-free interprocess communication in shared-memory real-time systems", Proc. 2nd UK Asynchronous Forum, University of Newcastle upon Tyne, Newcastle upon Tyne, UK, July 1-2, 1997.

[11]   Semenov, A. and Yakovlev, A., "Contextual net unfolding and asynchronous system verification", Technical report, No.572, Department of Computing Science, University of Newcastle, April, 1997.

[12]   Clark, I.G., PhD thesis, King's College, University of London, in preparation.

[13]   Xia, F., Combining MASCOT with Petri nets, PhD thesis, King's College, University of London, to be published in 1998.

[14]   Clark, I.G., Xia, F., Yakovlev, A.V. and Davies, A.C., "Petri net models of latch metastability", Electronics Letters, pp.635-636, Vol.34, No.7, April 2, 1998.

[15]   Jensen, K., Coloured Petri Nets. Basic Concepts, Analysis Methods and Practical Use. Volume 1, Basic Concepts. Monographs in Theoretical Computer Science, Springer-Verlag, 1992.

Chapter 3

PETRI NET REPRESENTATIONS OF COMPUTATIONAL AND COMMUNICATION OPERATORS

David H. Schaefer
Department of Electrical and Computer Engineering
George Mason University
Fairfax, VA 22030, USA
schaefer@gmu.edu

James A. Sosa
TASC, Inc
Chantilly, VA 20151, USA
jasosa@tasc.com

Abstract: This paper presents a collection of basic Petri net operators that are being utilized to model communication networks and parallel computing systems. The Petri net models contain two diagrams: the *Data Structure* which shows paths that are available for the transfer of data, and the *Control Structure* which specifies the control needed to manage the flow of data through the paths of the data structure. This paper will concentrate on descriptions of data structures. A concise abbreviated Petri net representation has been developed that allows for the portrayal of larger systems than would be possible if complete Petri net diagrams were drawn.

Keywords: Communication models, control structures, data structures, parallel computer architectures, Petri nets

A. Yakovlev et al.(eds.), Hardware Design and Petri Nets, 51-74.
© 2000 *Kluwer Academic Publishers.*

INTRODUCTION

The basic motivation for this work has been pedagogical. There is no standardized method of modeling computer and communication network architectures. Computer logic diagrams and schematic diagrams are much too detailed to present concepts involving complete systems.

The use of colored timed Petri nets provides a powerful method of distilling the basic essence of parallel computing concepts.

As this work has progressed there has been interest in the use of Petri net operators and the abbreviated Petri net versions for the specification of systems. The ability to determine the correctness of specifications through software execution of the models adds to both the specification and pedagogical usefulness.

The representation of a single operator, such as an individual adder with its input and output registers, will first be examined. The concepts developed will then be extended to operators that contain an array of computational elements (such as an array of adders or an array of buses). Finally the question of how to represent operators that allow communication between registers at distant locations will be addressed.

1.   COMPUTATIONAL AND COMMUNICATION OPERATORS

1.1   Nomenclature

The notation used by Aarhus University's *Design CPN* program will be generally followed. An exception is that transitions that have any arcs with timing notations are identified by rectangles and are referred to as *rectangular transitions*. Those that do not have any arcs with timing notations are represented as a line and are referred to as *line transitions*.

1.2   Generalized characterization

In Petri net representations of computation and communication systems:

a)   An operator is represented by a rectangular transition (in one case a rectangular transition and a line transition in series) with input and output data arcs, input and output control arcs, replacement arcs, and erasure arcs. These arcs are defined in Figure 3.1.

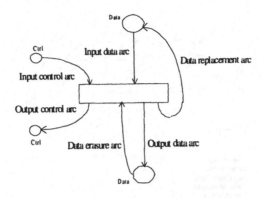

Figure 3.1 Arc Notation

b) The control input place of an operator transition represent ports through which tokens arrive from a control structure. The control output place represent ports through which tokens are returned to a control structure

c) Data places represent storage elements (registers, arrays of registers, memories and even a long wire that is storing information). The data replacement arc's function is to instantly replace the operand token destroyed when the transition fires. The function of the data erasure arc is to remove old result data, data that is being replaced by a new result. The output data arc always contains time notation resulting in the generation of inactive tokens in the output data place. The length of time the token remains inactive represents the time required for the output register to stabilize.

d) The value of data stored in a storage element is indicated by the numerical value of the v field of tokens in the place.

e) The node position of data stored in an array of storage elements is indicated by the value in the p field of tokens in the place.

f) The Petri representation of a master-slave register requires that two registers be represented. An operator *OP* loads a master register $X(m)$ with a result . Later an identity operator *I* loads the slave *register X(s)* with this result.

1.3   Operator representations

A single generalized computational operator is shown in Figure 3.2. An example is an adder (Figure 3.3). The *42* in place *S* (Figure 3.3(a)) is a token "left over" from a previous calculation   A r (red) cl control token resides in place *CL*.

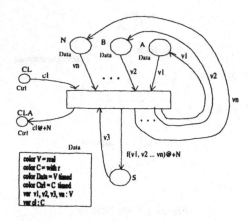

Figure 3.2 Single computational operator representation

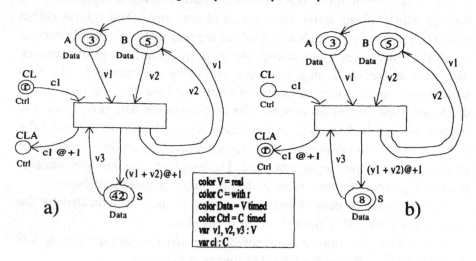

Figure 3.3 Addition operator. a) before transition firing. b) after transition firing

An array of logic or arithmetic elements, such as the inverters in Figure 3.4, is represented by a single transition. The register array that provides input to this operator is represented as a single place containing tokens whose values are number pairs, p,v. (Figure 3.5), where p represents a node location (position), and v the value of data stored at that location.

An array of registers connected to another array of registers through inverters therefore has the Petri representation of Figure 3.6. A previous calculation has set all the flip-flops in array B. Because control tokens for all positions are present, the transition is enabled for all values of p.

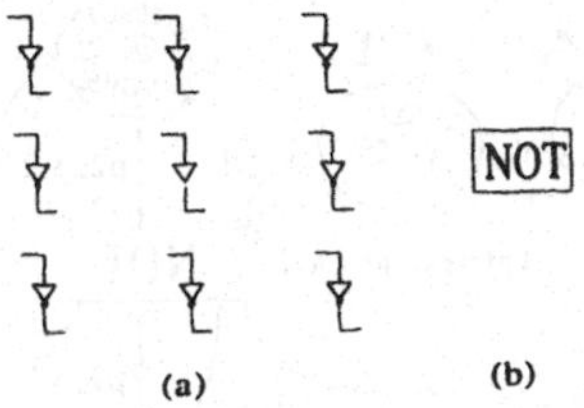

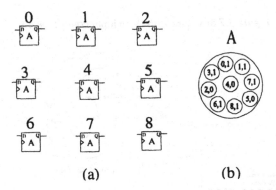

Figure 3.4 An array of Inverters (a) and (b) its represéntation by a single transition

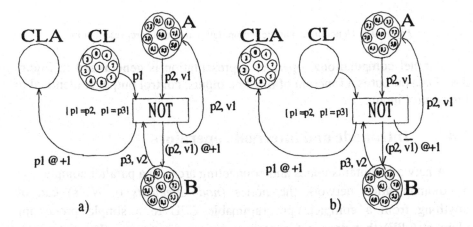

Figure 3.5 The registers in the array are represented by the tokens in the place

Figure 3.6 Parallel registers and operators a) before transition firing b) after transition firing

In contrast, in the computational structure shown in Figure 3.7 only the data in place *B* at positions one, four and seven are changed by the NOT operator. All other positions retain their original values.

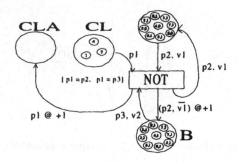

Figure 3.7 Structure with three input control tokens

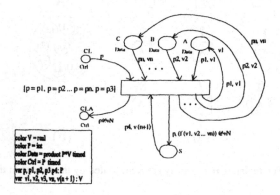

Figure 3.8 Generalized parallel computational operator representation

Parallel computational operator representation is generalized in Figure 3.8. Data output is a function of the data inputs, control output is identical to the control input.

1.4   Intranode and internode operators

A network contains nodes and connecting arcs. In a parallel computer or a communication network the nodes *(node elements* or *NE*'s) can be anything from a complete programmable CPU to a simple processing element (PE) that does not contain a program memory. The term *node element* will be used to embrace every type of node architecture.

We distinguish between two major types of operators, operators that operate within a node, *intranode (*or *within node) operators,* and those that provide for operation between nodes, *internode operators.* Internode operators will often be referred to as *between node* operators, or simply as *BN* operators. Similarly intranode operators are *within node* operators, or *WN* operators.

In addition to being either intranode or internode, operators can be either *computational* or *communication*.   An   array of adders is a WN computational operator, as each adder operates within a node. Busses that provide communication between registers inside computing elements together constitute a WN communication operator as each bus only exists within a node.

In contrast if registers located in different nodes provide the inputs to an OR circuit, then this OR circuit is a BN computational operator. A bus that provides communication between registers located at various nodes is an example of a BN communication operator. Neither a parallel computer nor a communication network can exist without internode communication operators.

We can therefore have an intranode bus (WN-BUS) operator or an internode bus (BN-BUS) operator.

1.5   Intranode (WN) communication operators

Let us examine a bus (Figure 3.9) that provides communication between registers *A, B, C* and *D* of a single computer. A control system provides switching signals to tristates and clock signals to the registers for data transfers. In the Petri representation (Figure 3.10) a control token specifies the register *sr* that is the source of the information to be transferred, and *dr*, the destination register into which the data is to be deposited. The rather extensive arc markings containing if statements define the operator.

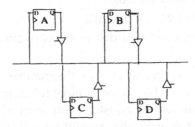

Figure 3.9 Single WN bus

As any of the registers can be either a source or a destination register, individual arcs do double duty. When the register is a source register the arc going into the transition is the data input arc. When the register is a destination register this same arc becomes an erasure arc. In a similar manner the arc from the transition to the place serves double duty as both the output data arc and the replacement arc. There can be more than one control token, but the *sr* must be the same for each one.

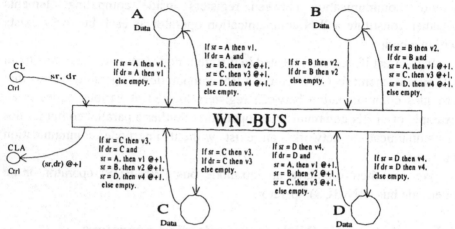

Figure 3.10 Petri repesentation of bus of Figure 3.9

The same diagram also represents a WN crossbar. Here there can be many control tokens, but each must have a unique *dr* field. In the Petri net representation the distinction between a bus and crossbar is contained in the control structure where the enforcement of rules concerning allowable control tokens takes place.

The general rule to change a single WN operator representation into parallel WN operator representation is to:

a) add a *p* field to all data variables

b) provide a *p* field for control tokens.

c) add a guard to the transition that restricts operation to the *p* in the control token field.

Applying these rules to the single bus diagram (Figure 3.10) results in Figure 3.11. For an array of buses the control token must not only specify the source *sr* and destination *dr* registers, but must also specify the node position *p* of the bus in the array that should transfer the data. If all buses are to transfer data at the same time then a control token for each position is required. As before there can be more than one control token for a given *p*, but the *sr* must be the same for each one.

This involved diagram can be simplified if we change the circuitry so that there is a set of source registers, and a set of destination registers, (Figure 3.12). Here any *A* register is able to send data to any *B* register within the node. In the Petri net representation (Figure 3.13), all the source registers of the network are lumped into one place, and all the destination registers into another.

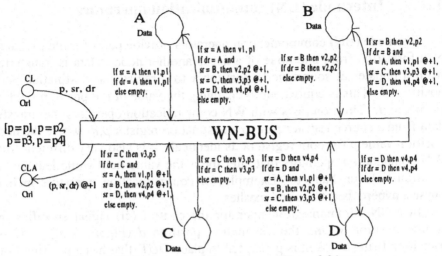

Figure 3.11 Petri representation of array of Figure 3.9 buses

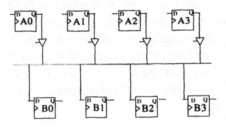

Figure 3.12 Bus with source registers, A, and destination registers, B

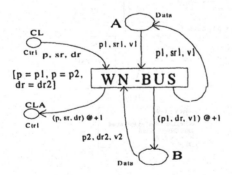

Figure 3.13 Array of Figure 3.12 buses

If, on the other hand, the operator is a cross-bar then values stored in any *A* registers can be transferred to any *B* register at the same time. Place *CL* can contain a control token for every source register, the only restriction being that the destinations must be unique.

1.6   Internode (BN) communication operators

Internode (BN) communication operators provide paths for the exchange of information from one node of a net to another node. Data is transferred from a register at source node position s to a register at destination node position d. This is equivalent to changing the value of the position field, p, from s to d. This contrasts with WN communication operators that transfer data from a source register p,sr to a destination register p,dr where the node position value p for both registers is identical. It is also very different from WN computation operators that alter the v, the value field while leaving the position field unchanged. Examples of communication operators include busses, hypercubes, trees and meshes.

In a BN communication operator the control (cl) signal specifies the source position s, and the destination position d (Figure 3.14). Upon transition firing a token is generated in place *OUT* that has a position field of d, and a value field that is the value associated with position s in place *IN*. The guard makes sure that the token with position field d in place *OUT* is destroyed to make room for its incoming replacement. The value in the input register remains unchanged, and the output register stabilizes in N time units. Two acknowledge tokens are generated in place *CLA*, one with value s (that informs the source that the transmission of the data is complete) and the other with value d (that informs the destination that data has been placed in a register at its location).

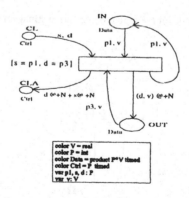

Figure 3.14 Internode (BN) communication operator

In Figure 3.15 the communication operator is a bus that provides communication paths that allow A registers at any of four locations to send data to B registers at the same four locations. A control system provides signals that enables the tristate at position s, and provides clock pulses at position *d*. Only one tristate can be enabled at any one time, but any combination of clock pulses can be sent to the *B* registers. This means that place *CL* is permitted to contain as many as four control tokens, but they must all have identical *s* fields.

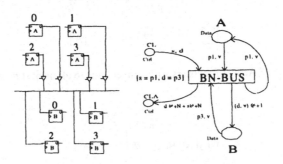

Figure 3.15. Bus BN communication operator (numbers over registers denode their location)

In Figure 3.16 if *s* equals *(x, y)* and *d* equals *(x+1, y)* (where *x* increases from left to right) then the transition represents what we call a *slider*, in this case an *eastern slider*. The registers in the *A* register array are connected to *B* array neighbors that reside in computing elements located to the east. It should be noted that no *d* notation is needed on the *CL* arc as *d* is determined by the value of *s*. Slider operators are used to represent mesh and toroidal topologies.

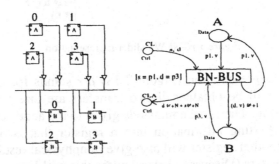

Figure 3.16 Eastern Slider BN communication operator

A composite slider, a NEWS operator (Figure 3.17) that can be configured to be either a northern, eastern, western or southern slider, allows data to be sent to any nearest neighbor. In the figure, the *A* registers are both the input and output registers.

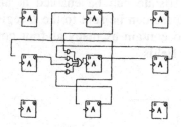

Figure 3.17 Array of registers with connections to all nearest neighbors

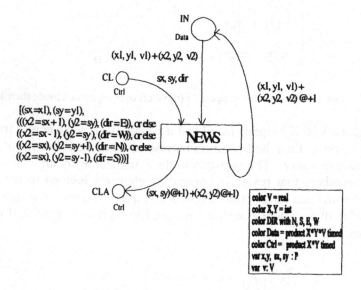

Figure 3.18 NEWS slider representation

In the Petri representation (Figure 3.18) *dir* stands for *direction*. Non physical results will be obtained if two transfers are directed to the same destination location. The same commands given to the actual circuit is likely to burn it up. Putting information into a register that is transferring its information to another register will also give nonphysical results. The use of master-slave registers (Discussed below) cures this problem.

Many types of communication operators exist in parallel systems, but, as we shall see, simple sliders are quite common. A NEWS operator forms the very common mesh topology.

1.7   Irregular pattern communication operator

Figure 3.19 shows a communication operator with an irregular connection pattern. For every source register there is a single destination register. Therefore a table that assigns a unique *d* to each *s* can be constructed and used as a part of arc markings. There is little use for such irregular interconnections paths in computers, but great use for them in communication systems.

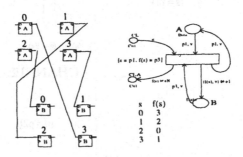

Figure 3.19 Irregular pattern operator

In all the systems we have discussed, new data can not be loaded into an input register until the output register has stabilized with its new value. This assures that the input to an operator doesn't change while its output register is loading.   Receipt of a cla acknowledgment signal by an external source indicates that the output has stabilized. It is then safe to reload the input registers.

It has been assumed that the four locations shown in Figure 3.19 are in one cabinet. If, however, they are far removed from each other the rules must change. For instance, if registers at location 1 are in New York, and the registers at location 2 are in Los Angeles, the connecting wire becomes a storage element itself. New York can reload its register thousands of times before Los Angeles has even received the first bit of a transmitted message. There is no fear that register loading in California will be disturbed by changes in state of the New York register.

The model for representing a communication channel operator (Figure 3.20) therefore includes three places, one representing the input registers at the source locations, the second the storage in the physical links, and the third the output registers at the destination locations. Two transitions are needed, a rectangular transition to generate the timed tokens, and a line transition to load the output place. The function f1(s) is the f(s) function from the previous example that specifies the destination of the link.

The rectangular transition places inactive tokens into the channel place. The length of time that they are inactive is f2 time units. Function f2, a function of the distance between source and destination, is the latency of the link.

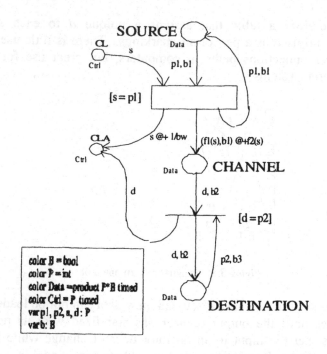

Figure 3.20 Communication channel operator

The inactive tokens in the channel place represent bits in transit. The maximum number of tokens that can be in the *CHANNEL* place for a given source is the maximum number of bits that can be held by the link. Note that only inactive tokens can reside in the *CHANNEL* place. As soon as they become active the line transition is enabled, and they are destroyed.

After the rectangular transition fires the *s* acknowledgment token generated in place *CLA* will be inactive for a length of time proportional to the latency. This means that an acknowledgment is supplied to the source at the bandwidth rate. The source is therefore able to supply new data at this rate.

Upon firing of the line transition old data at the destination position is destroyed, due to the feedback arc, and it is replaced by the data that has just been received (become active). Also, the *d* acknowledgment is generated in place *CLA*. The destination site therefore receives delayed ready tokens at the bandwidth rate. It should be emphasized that Figure 3.20 represents a collection of communication links, all between different locations, but all having the same bandwidth.

2. ABBREVIATED PETRI NET (APN) NOTATION

Abbreviated Petri nets are concise representations of full Petri net computational structures. Different techniques are used to represent data structures and control structures. In the representation of data structures in abbreviated Petri net (apn) notation:

a) Transitions of the full Petri net are duplicated (represented as rectangles) in the apn diagram. An exception is the identity transition of master-slave registers (see below).

b) If there is no node position field in the arc markings of the data input arc, then the data input place and its data input arc (together) are represented by a single line that is attached to the top of the transition.

c). If there is a position field in the arc markings of the data input arc, then the data input place and its data input arc (together) are represented by a double line that is attached to the top of the transition.

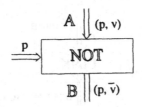

Figure 3.21 The apn representation of the Petri nets of Figure 3.6

d) Data output places and data output arcs (together) are represented by either single or double lines that are attached to the bottom of the transition. The use of single or double lines depends on the presence or absence of position fields in the arc markings

e) The existence of data replacement arcs and data erasure arcs are always assumed to be present, but are never represented.

f) The input control place and the input control arc are together represented by either a single or double line that is attached to the left side of the transition. The output control arc and the output control place are always assumed present, but never represented. The use of single or double lines depends on the presence or absence of position fields in the arc markings. The whole collection of control input and output places and arcs is known as a *control link*. Representing only the input control place and its arc effectively represents the link.

g) Arcs can be marked with generic markings.

When master-slave registers are present the shorthand shown in Figure 3.22 is used to make the apn diagrams even more abbreviated. The operator transition and the identity transition of the master-slave flip-flop are represented as one rectangle with two control links, as shown at the far right in the figure. Control tokens must be provided for both links. The apn representation of a communication channel (Figure 3.20) is the rectangle and line of Figure 3.22.

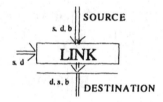

Figure 3.22 Master-slave representation

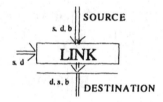

Figure 3.23 Communication channel apn representation

2.1   Masking

In SIMD computers the control structure produces identical control tokens for each element of the parallel operators. In most SIMD computers, however, there are operators that allow specified elements to ignore the global command. These ignoring elements, left with their old data, they are said to be *masked*. A *masking register* (G) specifies which elements are masked.

In the circuitry for carrying out this operation (Figure 3.24) B will be clocked only at those locations where G has a *one*. At these locations the B register is loaded. At locations where G is *zero* there is no clocking due to the action of the AND circuit and there is no loading of the B register. The arc notation for token b in the Petri representation of the masked NOT operator (Figure 3.25) formally states this function. In the apn notation the masking register appears as a small capitol letter over the control inputs.

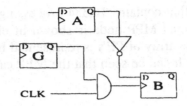

Figure 3.24 Masking circuitry

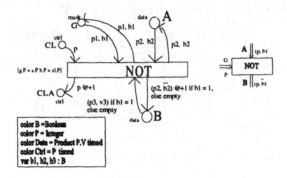

Figure 3.25 Petri representation of masking

2.2   The Massively Parallel Processor

The Massively Parallel Processor (MPP) (delivered to the Goddard Flight Center in 1983 and in operation until 1990)

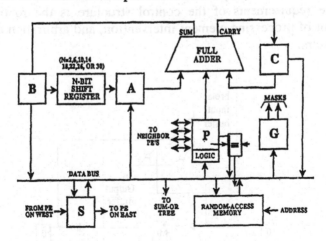

Figure 3.26 One processing element of the MPP

is a SIMD computer that contains 16,384 processing elements in a 128 by 128 array. An individual MPP node is shown in block diagram form in Figure 3.26. The whole array of PE's is represented by the computer's data structure (Figure 3.27). It can be seen that the MPP can be represented with five operators.

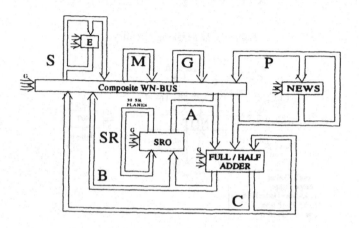

Figure 3.27 The Massively Parallel Processor

2.3   Cray T3E

A one dimensional version of the data structure of the SGI/Cray T3E MIMD computer (Figure 3.29) shows registers internal to the Processing Element *R*, the cache *C*, the E registers *E*, memory *M*, and the internodal communication registers *Q*. The circuitry of the *BN-E-XBAR* is shown in Figure 3.28. Among the requirements of the control structure is the routing of data independent of processing element intervention, and arbitration of access to the Q registers.

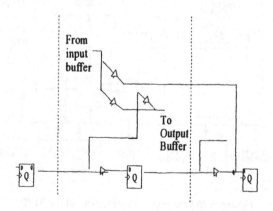

Figure 3.28 Slice of BN-E-XBAR operator

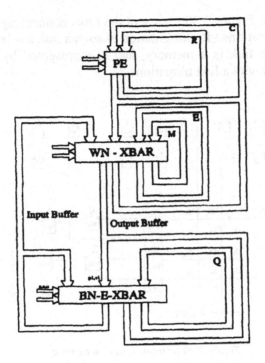

Figure 3.29 Simplified T3E type system

3.   SOME INTERNET CONCEPTS

Consider the data structure required to send a bit of information from one of a set of registers at location *p* to one of a set of distant registers (Figure 3.30). At each location are a set of gates, *g*, (like at an airport) from which data can be sent. To be specific a data structure is needed that transmits the data stored in register *A* at gate 2 (Lets call it register *A2*)located at position zero to register *B0* located either at distant location one, distant location two or distant location three at gate zero (the *0* in register *B0* is both the gate number and the sending location. (The diagram even allows data to be sent to register *B0* at location zero, the same location as the *A* register, but this is not an interesting case). A control system provides signals to the tristates of the crossbar, and clock signals to the registers that carry out the transfer.

The data from location zero must pass through interface registers *F0n,* where *n* is the desired destination location. In our case *n* can be either *1, 2* or *3*. It is then sent on to register *B0* at the distant location. We can generalize this to say that data from register *Ag* at source location *s* must pass through interface registers *Fsd,* and on to register *Bs* at the destination location.

The dotted lines in Figure 3.30 represent the links connecting location zero with the distant locations. Representation of such a link has been addressed. Recall that such a link is a memory, and is represented by a rectangular transition in series with a line transition.

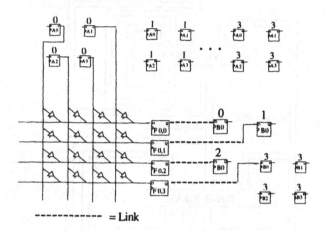

Figure 3.30 Internet concept schematic

The data structure required (Figure 3.31) shows the array of internal crossbars with a control signal specifying the source position, gate and destination position. The crossbar operator loads the indicated data to interface register (s,d). The link operator then needs only a *source, destination* command. After a time equal to one over the bandwidth of the link, the source receives an acknowledgment signal, and can reload its (A,g) register. After a length of time equal to the latency of the link, the destination will receive a ready signal and loads its (B,d) register. The data structure just presented is general and can be the model for millions of A and B registers spread around the world. It models a system where every location can be reached from every other location. In practice there is no such dense network of connections. This data structure is, however, a starting point. Its apn representation is shown in Figure 3.32.

The system described will deliver data from any A register to any B register if links from every location to every other location are present. If the link from position zero to position three does not exist, for instance, Figure 3.32 must be altered if we wish to send data from register $A2$ at position zero to a register at position three. If all other links are present we can first send the data to register $B0$ at position two, transfer it to register $A0$ at position two, and from there send it on to register $B2$ at position three.

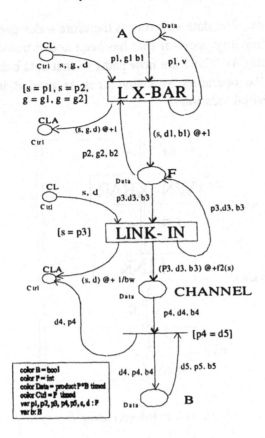

Figure 3.31 Data structure for Figure 3.30

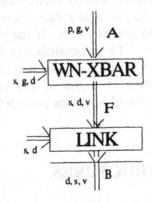

Figure 3.32 apn representation of Figure 3.31

In order to accomplish this pair of transfers an operator that allows the transmission of data from *A* registers to *B* registers is required. An identity operator is adequate. The data structure is therefore redesigned as shown in Figure 3.33. The identity operator that has been added transfers data from register *Bs* to register *As*. There are now paths so that data can go round and round traversing the operators as many times as needed in order to be delivered to the desired location

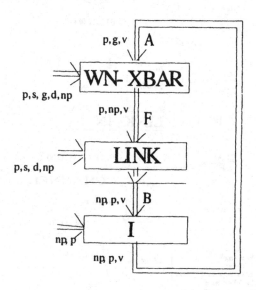

Figure 3.33 Indirect routing

The control must now be able to a) look up the location to send data to best get it closer to its destination, and b) recirculate the control tokens. In our example the command is, send any data whose position is zero, and whose destination is three to position two. If the destination is any other position send it straight there. To accomplish this the control tokens must have five fields: p, *s*, *g*, *d* and *np* (for *next position*). The p field is needed as the position of the data when in transit is not at either the source or destination locations. The control structure provides the tokens that steers the data to its destination.

4.   CONTROL STRUCTURES

A data structure represents hardware. Control signals make the hardware come alive.   or destination locations. A *control structure* provides the needed control inputs to a data structure by delivering *configuration and load* (cl) signals to the data structure. In turn the control structure receives acknowledge signals (called *configuration and load acknowledgments* or *cla signals*) from the data structure. Acknowledge signals are necessary as a control system must be made aware of when the loading of registers is complete.

A detailed discussion of control structures is beyond the scope of this paper. In brief, a control structure consists of control modules which forward control tokens, transform control tokens, and even act as policeman allowing certain tokens to proceed while holding others back.

An example of a token transforming control module is the forwarding table module (Figure 3.34) used in the control structure that provides control tokens to the circuitry of Figure 3.31 where the link from location zero to link three does not exist. The module will generate a token in place *C2* that has a *np* field that will be two if the position is zero and the destination is three.

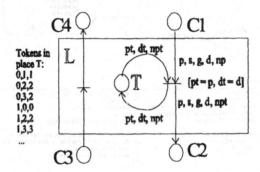

Figure 3.34 Forwarding table control module

5.   DESIGN CPN USE

Use is made of software tools to aid in instruction and to check the accuracy of the models. An example of the use of Design CPN is a model of a bus feeding a master slave register (Figure 3.35).

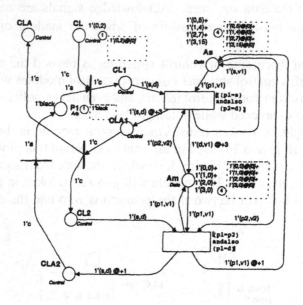

Figure 3.35 Design CPN model of a bus feeding master-slave registers

ACKNOWLEDGMENT

The authors would like to acknowledge the helpful suggestions of Douglas Lamb.

REFERENCES

[1] Jensen, K., *Coloured Petri Nets*, Springer-Verlag, Berlin, 1992.

[2] Allen, J., Gallager, R., *Computation Schemata and Implementations* (unpublished notes), Mass. Inst. of Technology, 1982.

[3] Schaefer, D.H. *The Characterization and Representation of Massively Parallel Computing Structures*, Proceedings of the IEEE, April 1991.

[4] Shapiro, R. M. *Validation of a VLSI Chip Using Hierarchical Coloured Petri Nets.* Journal of Microelectronics and Reliability, Special Issue on Petri Nets, Pergamon Press, 1991.

[5] Parallel Computer Architecture - a Unified Approach, Schaefer. D.H. and Sosa, J.A. Research Studies Press (in preparation).

II
MODEL ANALYSIS AND VERIFICATION FOR ASYNCHRONOUS DESIGN

Chapter 4

PROPERTIES OF CHANGE DIAGRAMS

Uwe Schwiegelshohn

Computer Engineering Institute, University Dortmund
D-44221 Dortmund, Germany
uwe@carla.e-technik.uni-dortmund.de

Lothar Thiele

Computer Engineering and Networks Laboratory (TIK)
Swiss Federal Institute of Technology (ETH) Zurich
CH-8092 Zürich, Switzerland
thiele@tik.ee.ethz.ch

Abstract   The paper investigates properties of change diagrams. They are able to model a subclass of concurrent systems, for example asynchronous circuits or timing diagrams. The following results are described: Change diagrams are related to the class of dynamic min-max graphs. Efficient algorithms for timing analysis are derived. Liveness and boundedness properties are investigated.

Keywords: asynchronous circuits, change diagrams, dynamic min-max graphs, timing analysis, timed change diagrams

1.   INTRODUCTION

Change diagrams (CD) have been introduced and described in [18, 8, 9]. This model is characterized by the following properties, see also [21]:

- CDs can equally model two types of causality, i.e. AND causality and OR causality, see [21]. For example, if an event c has two cause events a and b, then event c occurs at most i times if the minimum of occurrences of a and b equals i (AND causality) or if the maximum of occurrences of a and b equals i (joint OR causality).

- CDs cannot directly represent processes with conflicts or choice.

Additional properties and a comparison to other models of computation as well as many examples for their application to the analysis and

77

A. Yakovlev et al.(eds.), Hardware Design and Petri Nets, 77-92.
© 2000 *Kluwer Academic Publishers.*

synthesis of circuits and protocols can be found in [9, 20, 21] and the references therein.

Initially, CDs have been introduced in a labeled and interpreted form in order to directly establish the link to signal transitions. In the present paper, we address the more general unlabeled version of the model.

Definition 1 (Untimed Change Diagram) *An untimed change diagram (CD) is a graph $G = (V, E)$ with a marking (or state) function $d : E \to \mathcal{Z}$ where $d(e)$ of edge $e \in E$ is called the number of tokens on e or the marking of e. The set of nodes V is partitioned into two sets V^+ and V^- ($V = V^+ \cup V^-$).*

In order to avoid notational difficulties we assume that each node has at least one incoming edge and that G is connected. However note that all results in this paper can be easily generalized.

Definition 2 (State Transition) *A node $v \in V$ is enabled at marking (or state) d iff*

$$d(u, v) > 0 \text{ for all } (u, v) \in E \qquad if v \in V^+$$

$$there\ exists\ (u, v) \in E \text{ with } d(u, v) > 0 \qquad if v \in V^-$$

An enabled node v may fire leading to a state transition with

$$d'(u, v) = d(u, v) - 1 \quad if (u, v) \in E, u \neq v$$

$$d'(v, u) = d(v, u) + 1 \quad if (v, u) \in E, u \neq v$$

All others markings remain unchanged in a state transition. The state transition is denoted as $d \xrightarrow{v} d'$.

At any moment any edge of G may contain either some negative ($d(e) < 0$) or some positive ($d(e) > 0$) tokens. Note that there is an immediate "recombination" of negative and positive tokens on any edge such that they cannot coexist on the same edge at any time. More complex transitions with concurrent firing of several vertices can always be converted into a sequence of simple transitions. We say that a state or marking d' can be reached from d if there is a sequence of state transitions σ leading from d to d', denoted as $d \xrightarrow{\sigma} d'$.

The well known class of marked graphs is a subclass of change diagrams characterized by $V^- = \emptyset$ and $d(e) \geq 0$ for all edges e.

In Figure 4.1a we show an example from [21] which corresponds to a low latency hardware structure with redundancy and buffers of length n. After firing of v_1, v_2, v_3, v_5 and v_6 3 times each, the state as shown in Figure 4.1b is obtained.

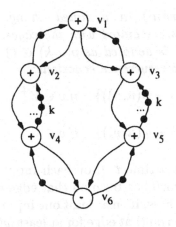
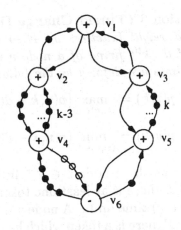

a) initial change diagram   b) change diagram after transitions

Figure 4.1 Change diagram corresponding to a redundant hardware system. The marking $d(e)$ is either shown as a number attached to e or as filled $(d(e) > 0)$ or empty $(d(e) < 0)$ circles.

In order to check the correctness of a specification or implementation, it is obviously desirable to derive properties of a given change diagram. In particular, because of the similarity to the model of marked graphs, it can be expected that comparable results can be obtained, see [5, 3]. Several results in this direction are already given in [9].

Until now, no general approach to characterize and determine liveness and boundedness properties of change diagrams is known. A CD is live for an initial marking d if from each reachable marking d' there exists for each node $v \in V$ a reachable marking d'' that enables v. A CD is bounded for an initial marking d if there exists an integer K such that for each reachable marking we have $|d(e)| \leq K$ for all edges $e \in \dot{E}$.

The paper contains the following new results concerning untimed change diagrams:

- A simple algorithm for deciding the liveness of a change diagram is given.

- The boundedness is characterized and an efficient algorithm is provided.

- Several properties of change diagrams are derived.

In order to evaluate system performance in case of an implementation or to verify timing properties of a system specification, timed change diagrams can be defined as follows:

Definition 3 (Timed Change Diagram) *In a timed change diagram, a weight function* $w : E \rightarrow Z$ *is associated with the edges. The time of the kth firing of a node $v \in V$ is denoted as $p(v, k) \in Q$. The firing times are defined by the following recurrence equations:*

$$p(v, k) = \max_{(u,v) \in E} \{p(u, k - d(u, v) + w(u, v))\} \quad if \, v \in V^+$$

$$p(v, k) = \min_{(u,v) \in E} \{p(u, k - d(u, v) + w(u, v))\} \quad if \, v \in V^-$$

In other words, a node $v \in V^+$ fires at a time t, if on each input edge $(u, v) \in E$ there is at least one token which has been on that edge for at least $w(u, v)$ time units. A node $v \in V^-$ fires, if on at least one input edge $(u, v) \in E$ there is a token which has been on that edge for at least $w(u, v)$ time units. This property can best be described with the unfolding of a change diagram. For each firing of a node $v \in V$ there is a new node in the unfolding. As the resulting graph is acyclic, the time of a firing can simply be calculated solving min- and max-equations for each node in the unfolding. An example is shown in Figure 7.2 where the ordered firing sequence is $((v_1, 0), (v_4, 0), (v_2, 1), (v_3, 1), (v_4, 1), (v_3, 2), (v_1, 3), (v_2, 4), ...)$. It can be seen, that the system consists of two bounded subsystems, i.e. v_1, v_2 and v_3, v_4, which evolve with the different firing periods 3 and 1, respectively. The two subsystems are connected via edges with an unbounded number of token. This behavior is called quasi-periodic in [16, 17].

The analysis of timing properties will be carried out by establishing a link between the class of timed change diagrams and dynamic min-max systems, see [11, 12, 1, 6, 7, 16, 17].

In particular, the present paper contains the following new results for the analysis of timed change diagrams:

- The first efficient (pseudo-polynomial) algorithm for the determination of the periods of a timed change diagram is derived.

- Results on the uniqueness of the periods are given.

- The boundedness of change diagrams is characterized and efficient (pseudo-polynomial) algorithms are derived.

2.   PROPERTIES OF UNTIMED CHANGE DIAGRAMS

2.1   BASIC PROPERTIES

Marked graphs are closely related to change diagrams. On the other hand, CD have two classes of nodes with different enabling conditions.

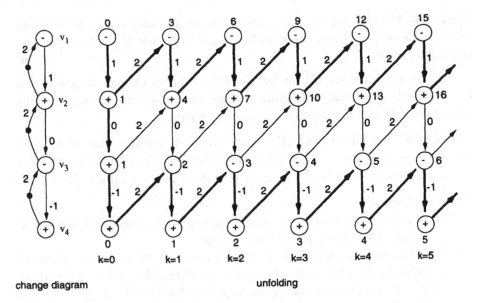

Figure 4.2 A change diagram and the corresponding unfolding. The edges of the change diagram and the unfolding are labeled with $w(e)$. The nodes of the unfolding are labeled with the associated firing time.

As a consequence, the markings can be negative both at the initial state or during state evolution. Therefore, some of the simple properties of marked graphs, see e.g. [5, 3], are no longer valid. Nevertheless, some properties can easily be derived.

Proposition 4 (Cycles) *Let C be a directed cycle of a change diagram with initial state d. For any reachable state d', the sum of tokens in C is constant, i.e.*

$$\sum_{e \in C} d(e) = \sum_{e \in C} d'(e)$$

Sketch of Proof: The proof as known for marked graphs only uses the firing rule and not the enabling conditions. Consequently, the result holds. ∎

The next property characterizes firing sequences which reproduce the state of a change diagram.

Proposition 5 (Reproducibility) *Let σ be a sequence of state transitions which reproduces a state, i.e. $d \xrightarrow{\sigma} d$. If the change diagram is connected, then each node $v \in V$ occurs in σ equally often.*

Sketch of Proof: Again, the proof as known for marked graphs uses the firing rules only and not the enabling conditions, see [5]. Consequently, the result holds. ∎

Finally, a simple condition for the liveness of a change diagram can be obtained. But first, a Lemma will be shown which stresses the importance of directed cycles in a change diagram.

Lemma 6 (Positiveness) *A change diagram with initial state d is given. If $\sum_{e \in C} d(e) > 0$ for all directed cycles C, then there exists a firing sequence σ with $d \xrightarrow{\sigma} d'$ such that $d'(e) \geq 0$ for all $e \in E$.*

Sketch of Proof:

The proof is constructive. In each step, the sum of negative markings, i.e. $\sum_{e \in E} \max\{0, -d(e)\}$ is reduced by one.

Let us choose one edge $(u, v) \in E$ with $d(u, v) \leq 0$. If the source node u can be fired, then the sum of negative markings has been reduced by one. If not, it has at least one input edge (w, u) with $d(w, u) \leq 0$. Again, if w can be fired, then the sum of negative markings has been reduced by one. Continuing this process, at least one node can be fired. Otherwise, a cycle with a non-positive sum of markings exists. ∎

Consequently, if the sum of tokens is strictly positive in any cycle, then there is a firing sequence which drives the change diagram into a "normalized" form where no marking is negative.

Proposition 7 (Liveness) *A change diagram with initial state d is given. If $\sum_{e \in C} d(e) > 0$ for all directed cycles C, the change diagram is live.*

Sketch of Proof: We have shown that there exists a firing sequence which yields $d(e) \geq 0$ for all edges $e \in E$. Now, if we restrict ourselves to a change diagram which contains nodes $v \in V^+$ only, then the graph is a marked graph and the same proof as e.g. in [5] holds.

If a change digram with $V = V^+$ is live, there exists a firing sequence such that any node can fire arbitrarily often. This property still holds if if some nodes change their type from $v \in V^+$ to $v \in V^-$.

It can be proven that a change diagram is live iff there is a firing sequence such that any node can fire arbitrarily often. This proof is based on the persistence of change diagrams. Therefore, the result of the Proposition holds.

Unfortunately, the above result gives a sufficient condition for the liveness of a change diagram only. In fact, Figure 4.3 shows the example of a live CD which violates the condition of the above Proposition. The next section contains also necessary conditions for liveness.

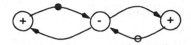

Figure 4.3 A live change diagram with a negative sum of markings in a directed cycle.

2.2   LIVENESS

Is a change diagram live for a given initial state d? We address this problem by presenting a constructive algorithm for its solution. A similar approach as in [17] is used. In particular, an algorithm "live" is defined which is particularly easy to analyze and which constructs a certain firing sequence. Note that we assumed at the beginning of the paper that each node has at least one incoming edge and that G is connected.

The algorithm uses the change diagram G and the initial state d and returns 'true' if G with d is live and 'false' otherwise. If it returns 'false', the graph G_t is a maximal live subgraph of G, i.e. the live subgraph with the maximal number of nodes.

Internally, it uses a 'potential' $p(v) \geq 0$ for any node $v \in V$ which denotes the number of firings of node v in the firing sequence constructed so far. Moreover, it uses an upper bound on the number of markings of any simple path in G:

$$s = \sum_{v \in V} \left(\max_{(u,v) \in E} \{|d(u,v)|\} \right)$$

This algorithm simulates the change diagram by assuming that a vertex fires whenever it is able to do so. More precisely, a vertex $v \in V$ will fire k-times in a state d with

$$k = \min_{(u,v) \in E} d(u,v) \quad \text{if } v \in V^+$$

$$k = \max_{(u,v) \in E} d(u,v) \quad \text{if } v \in V^-$$

provided $k > 0$ holds. The resulting sequence of states starting with the initial state d is called a *greedy* sequence. Note that the greedy sequence is equivalent to the sequence of the corresponding time CD with $w(e)) = 0$ for every edge e.

In order to proof the main result of this section, the firing sequences of change diagrams must be considered more closely.

Lemma 8 *A change diagram G with initial state d is live iff there exists a firing sequence which contains each node $v \in V$ at least once.*

Boolean Function live(G, d, G_t) {
      in G, d; out G_t
      $p(v) = 0$ for all $v \in V$;
  loop: **do in parallel** (for all $v_j \in V^+$) {
         $p(v_j) = \max\{0, \min\{p(v_i) + d(v_i, v_j) \mid (v_i, v_j) \in E\}\}$; }
      do in parallel (for all $v_i \in V^-$) {
         $p(v_j) = \max\{0, \max\{p(v_i) + d(v_i, v_j) \mid (v_i, v_j) \in E\}\}$; }
      if (there was no change in any $p(v)$ for all $v \in V$) {
         $G_t = \emptyset$; **return** 'false'; }
      if ($p(v) > 0$ for all $v \in V$) { $G_t = G$; **return** 'true'; }
      if (there were only changes in $p(v)$, $v \in V$ with $p(v) > 2s$) {
         $G_t =$ subgraph induced by nodes with $p(v) > s$;
         return 'false'; }
      goto loop;
}

Table 4.1 Algorithm *live*

Sketch of Proof: Let us suppose that there is no sequence which contains each node at least once. Then there is a node v which cannot be enabled and the change diagram is not live.

Let us suppose that a firing sequence σ exists which contains each node at least once, e.g. $d \xrightarrow{\sigma} d'$. Then one can simply construct sequences which fire each node at least k times for any given k. To this end, a sequence σ' is determined from σ which contains each node exactly once and reproduces d', i.e. $d \xrightarrow{\sigma} d' \xrightarrow{\sigma'} d'$. Obviously, σ' can be applied to d' infinitely often.

As change diagrams are persistent, this proves the liveness. The detailed proof will be contained in the full paper.   ∎

Now, the main results of this section can be described.

Proposition 9 (Liveness) *Given a change diagram G with initial state d. The change diagram is live iff algorithm* live *returns 'true'.*

Sketch of Proof: At first note that $p(v)$ does not decrease for any node during the execution of the loop. Therefore, the function definitely terminates.

Because of the previous Lemma, the change diagram is live if the algorithm returns 'true'.

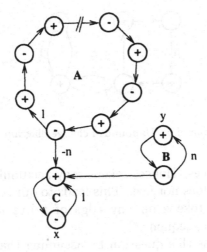

Figure 4.4 Change diagram.

Moreover, a change diagram is not live if no node can fire. In addition, if only nodes with $p(v) > s$ fire in an iteration, no node with $p(v) \leq s$ will fire in further iterations as a firing of nodes with $p(v) > s$ cannot lead to an enabling. ∎

Up to $|V| \max_{(v_i,v_j)\in E}\{|d(v_i, v_j)|\}$ iterations of Algorithm 4.1 may be necessary before it can be concluded that a min-max Petri net is live. As vertices may fire more than once per iteration, some vertices may be required to fire up to $O(|V| \max_{(v_i,v_j)\in E}\{|d(v_i, v_j)|\})^2$ times. An example for such a case is given in Figure 4.4. In our figures any vertex of V^+ is denoted by a small + sign inside the vertex circle. A similar convention is used for the vertices from V^-. If cycle A consists of k vertices, vertex x will only fire at iteration $kn + 2$ of Algorithm *live*. At this time vertex y has already fired $\lceil \frac{kn^2+2n}{2} \rceil$ times.

Note that several simple improvements of Algorithm 'live' are possible. In addition, the established relationship between the liveness of change diagrams and the static min-max problem also allows the use of more efficient algorithms like those described in [16]. Those algorithms determine after k iterations that Cycle A in Figure 4.4 is independent of any other vertex. Then the potential of all vertices of Cycle A is set to infinity, thus causing vertex x to fire far earlier.

2.3   BOUNDEDNESS

Next assume that we are given a live min-max Petri net. Note that in general there are no timing constraints which uniquely determine the

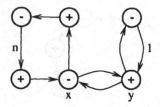

Figure 4.5   Unbounded change diagram.

fire sequence of vertices. Hence, tokens may accumulate on some edges as the target vertex does not fire. This leads to our second key problem.

Is the number of tokens on any edge of a live min-max Petri net bounded for any firing sequence?

We start to address this question by assuming that the graph of the change diagram is strongly connected.

Proposition 10 *A live change diagram is unbounded if the underlying graph is not strongly connected.*

Sketch of Proof: Let us suppose that the graph is not strongly connected. Then a tree of strongly connected subgraphs can be formed and an unbounded number of tokens will accumulate on some edge if no vertex of a leaf component is fired. ∎

But an unbounded accumulation of tokens on some edge may not only be due to the refusal to fire some vertices. In Figure 4.5 a simple live change diagram is given where even the greedy sequence results in an accumulation of positive tokens on edge (x, y) and a corresponding accumulation of negative tokens on edge (y, x) if $n > 2$.

The next lemma gives a necessary and sufficient condition for the existence of sequences with an unbounded number of tokens on some edges.

Lemma 11 (Boundedness) *Assume a live change diagram with initial state d based on a graph $G = (V, E)$. There are sequences causing an accumulation of an unbounded number of tokens on some edge if V can be partitioned into V_1 and V_2 such that*

1. *V_1 is a live change diagram when only edges between nodes of V_1 are considered.*

2. *There are no edges $(u, v) \in E$ with $u \in V_2$ and $v \in V_1 \cap V^+$.*

Sketch of Proof: Let us suppose that there exists a firing sequence which leads to an number of markings $|d'(e)| > K$ with

$$K = \sum_{(u,v) \in E} |d(u,v)|$$

in some edges. These edges will be called unbounded edges. If a firing sequence leads to an unbounded edge, G with initial state d is unbounded, see the proof of correctness for algorithm 'live'. By cutting all unbounded edges the resulting graph is not connected any more. Therefore, we can partition V according to $V = V_1 \cup V_2$ such that V_1 contains all sources of unbounded edges with $d(e) > 0$, all targets of unbounded edges with $d(e) < 0$ and all nodes which are connected to them via bounded edges. The nodes in V_1 can fire arbitrarily often and independent of those in V_2 as the connecting edges are unbounded, and the nodes in V_1 fired already at most K times more often than those in V_2. Edges $(u,v) \in E$ with $v \in V_1 \cap V^+$ cannot exists as otherwise, $v \in V_1$ would not have fired sufficiently often. Therefore the partitioning in the Lemma exists.

Let us suppose that the partitioning exists, then the subgraph induced by nodes V_1 (without edges between V_1 and V_2) is live. There exists a sequence of firings which contains all nodes in V_1 arbitrarily often and nodes in V_2 a finite number of times as there are no edges $(u,v) \in E$ with $v \in V_1 \cap V^+$. This sequence leads to unbounded edges which connect V_1 and V_2. ∎

Unfortunately, Lemma 11 by itself does not guarantee an efficient algorithm. Although Algorithm *live* allows the separation of dead vertices from live ones, there remains the problem of selecting those edges which must be deleted to perform the partition.

The following algorithm *bounded* returns 'true' if the CD is bounded and 'false' otherwise. It simply tests the liveness of the change diagram after inhibiting the firing of each of its nodes one after the other.

Proposition 12 *A given connected change diagram G with initial state d is bounded iff algorithm* bounded *returns 'true'.*

Sketch of Proof: The algorithm obviously terminates. Let us suppose that *bounded* returns a subgraph G_t. Obviously, there exists a firing sequence which contains all nodes in this subgraph arbitrarily often and nodes not in G_t a bounded number of times. As the change diagram is connected, there is at least one edge connecting a node in G_t and one not in G_t. The marking on this edge would be unbounded.

Now, let us suppose that the change diagram is not bounded. Then according to Lemma 11 there exists a partition $V = V_1 \cup V_2$ where the

```
Boolean Function bounded(G, d) {
    in G, d;
    for each v ∈ V do
        G' = G;
        remove all edges (u, v) from G';
        live(G,d,Gₜ);
        if ( Gₜ ≠ ∅ ) return 'false';
    return 'true';
}
```

Table 4.2   Algorithm *bounded*

subgraph induced by V_1 is live. Consequently, if we prevent any of the nodes $v \in V$ from firing, the subgraph induced will still be live, the algorithm 'live' returns $G_t \supseteq V_1$ and Function 'bounded' returns 'false'.

∎

3.   PROPERTIES OF TIMED CHANGE DIAGRAMS

The analysis of timed change diagrams, see e.g. Figure 4.2, is closely related to the area of interface timing verification, see [10, 19, 22]. In these problems, the solution to a set of equations involving min and max operators is required. But caused by the dynamic nature of change diagrams, a much more general problem must be solved here. On the other hand, for timed marked graphs corresponding results are known for a long time, see e.g. [4, 15, 14, 2, 1].

After several attempts to solve the analysis of dynamic min-max problems, see [11, 13, 1, 6, 7], the first efficient (pseudo-polynomial) algorithm for the determination of the periods of dynamic min-max systems has been described in [16, 17]. In addition, a more general class of dynamic min-max systems is considered and results on the uniqueness and existence of their periods are derived.

The results obtained in [16] and [17] hold for the class of timed change diagrams as well. Change diagrams are called static graphs and their unfoldings are called dynamic graphs. For the rest of this paper we suppose that $\sum_{(u,v) \in C} d(e) > 0$ and $\sum_{(u,v) \in C} w(e) > 0$ for all cycles C of the change diagram. As has been shown in the present paper, this property guarantees the liveness of the change diagram.

The following results from [16, 17] can be applied to the timing analysis of change diagrams.

Periodic Case Suppose that only periodic firing times of the nodes are considered, i.e.

$$p(v, k+1) = p(v, k) + \lambda \quad \text{for all } v \in V$$

where $\lambda \in Q$ is called the (common) period of the timed change diagram.

Then either the change diagram has no periodic firing times or the period is unique. Moreover, there is an algorithm which computes the period in pseudo-polynomial time.

The algorithm uses an function similar to 'live' and a binary search to determine the period of the change diagram.

Even if no common period of all nodes exists there are individual periods of the nodes. This property is called quasi-periodicity.

Quasi-Periodic Case Suppose that an individual period is associated with each node of the change diagram, i.e.

$$p(v, k+1) = p(v, k) + \lambda(v) \quad \text{for all } v \in V$$

where $\lambda(v) \in Q$ is called the period of node $v \in V$. An example for this situation is shown in Figure 4.2. The periods of the nodes satisfy $\lambda(v_1) = \lambda(v_2) = 3$ and $\lambda(v_3) = \lambda(v_4) = 1$.

It can be derived from the results in [16, 17] that any timed change diagram has quasi-periodic firing times as already mentioned above. Moreover, the periods $\lambda(v)$ for all $v \in V$ are unique. Again, a pseudo-polynomial algorithm is given to determine these periods.

These results can be applied to determine the period (or cycle time, throughput) of asynchronous circuits which can be modeled with change diagrams. In addition, the throughput of iterative protocols involving AND and OR causality can be calculated.

The results shown in [16] and [17] apply to specific initial (first) firing times of the nodes, namely those which lead to a periodic or quasi-periodic firing. But very often one is interested in the firing times under arbitrary initial conditions. Though it is an open problem whether in this case the firing times become eventually periodic in a steady state, useful bounds on the firing times are derived in the present paper. As a consequence, a strong statement on the boundedness of timed change diagrams can be made.

Proposition 13 (Firing Times) *Given a timed change diagram G with initial state d and weights w.*

Then there exist constants k_1 and k_2 such that the firing times $p(v, k)$ of a node $v \in V$ can be bounded by

$$k\lambda(v) + k_1 \leq p(v, k) \leq k\lambda(v) + k_2$$

As a direct consequence of the above Proposition, the periods $\lambda(v)$ are unique and can be computed in pseudo-polynomial time with the algorithm given in [16, 17]. Intuitively, the nodes v fire with different 'speeds' determined by $\lambda(v)$.

Finally it is interesting to determine the boundedness of a timed change diagram. Of course, any bounded change diagram is bounded also in its timed version. On the other hand, as the timing restricts the set of possible firing sequences, the converse is not true. For example, a structurally not bounded change diagram may be bounded if it is executed with an appropriate timing.

Considering the example shown in Figure 4.2, it is obvious that the edges connecting the nodes v_2 and v_3 are unbounded. This observation can be generalized as follows.

Proposition 14 (Boundedness) *Given a timed change diagram G with initial state d, weights w and periods $\lambda(v)$ for all $v \in V$. The change diagram is bounded iff all periods are equal.*

Sketch of Proof: The difference of firing times between two nodes u and v satisfies

$$|p(v, k) - p(u, k)| \leq k|\lambda(v) - \lambda(u)| + 2s$$

If all periods are equal, the firing times of all nodes differ by a constant only. Therefore, all connecting edges have a bounded number of token.

After some calculations it can also be shown that at some time t the difference between the number of firings of u and v is larger than

$$k_1|\lambda(v) - \lambda(u)|t + k_2$$

for some constants $k_1 > 0$ and k_2. If there are nodes with different periods in G, then there exists at least one edge connecting two nodes with different periods. Remember that we consider connected change diagrams only. The difference in the number of firings grows with t which leads to an unbounded number of tokens on this edge.   ∎

4.   CONCLUDING REMARKS

The paper presents new results on the dynamic behavior of timed and untimed change diagrams. Major results concern the liveness and boundedness of timed and untimed CDs as well as a characterization of the firing times. These results can be applied to analyze properties of a certain class of concurrent systems.

It is an open problem whether timed change diagrams eventually show a periodic behavior independent of the initial firing times.

References

[1] Baccelli, F., Cohen, G., Olsder, G., and Quadrat, J.-P. (1992). *Synchronization and Linearity*. John Wiley, Sons, New York.

[2] Cohen, G., Dubois, D., Quadrat, J. P., and Viot, M. (1985). A linear-system-theoretic view of discrete-event processes and its use for performance evaluation in manufacturing. *IEEE Transactions on Automatic Control*, AC-30, No. 3:210–220.

[3] Commoner, F., Holt, A., Even, S., and Pnueli, A. (1971). Marked directed graphs. *Journla of Computer and System Sciences*, 5:511–523.

[4] Cunninghame-Green, R. (1962). Describing industrial processes and approximating their steady-state behaviour. *Opt. Res. Quart.*, 13:95–100.

[5] Desel, J. and Esparza, J. (1995). *Free Choice Petri Nets.* Number 40 in Cambridge Tracts in Theoretical Computer Science. Cambridge University Press.

[6] Gunawardena, J. (1993). Timing analysis of digital circuits and the theory of min-max functions. In *TAU'93, ACM International Workshop on Timing Issues in the Specification and Synthesis of Digital Systems*.

[7] Gunawardena, J. (1994). Min-max functions. Technical Report to be published in Discrete Event Dynamic Systems, Department of Computer Science, Stanford University, Stanford, CA 94305, USA.

[8] Kishinevsky, M., Kondratyev, A., and Taubin, A. (1991). Formal methods for self-timed design. In *Proc. EDAC 91*, pages 197–201.

[9] Kishinevsky, M., Kondratyev, A., Taubin, A., and Varshavsky, V. (1993). *Concurrent Hardware: The Theory and Practice of Self-Timed Design*. John Wiley and Sons.

[10] McMillan, K. and Dill, D. (1992). Algorithms for interface timing verification. In *IEEE Int. Conference on Computer Design*, pages 48–51.

[11] Olsder, G. (1991). Eigenvalues of dynamic max–min systems. *Discrete Event Dynamic Systems: Theory and Applications*, pages 1:177–207.

[12] Olsder, G. (1992). About difference equations, algebras and discrete events. *Topics in Engineering Mathematics*, pages 121–150.

[13] Olsder, G. (1993). Analyse de systèmes min-max. Technical Report 1904, Institut National de Recherche en Informatique et en Automatique.

[14] Ramamoorthy, C. (1980). Performance evaluation of asynchronous concurrent systems using Petri nets. *IEEE Transactions on Software Engineering*, pages 440–449.

[15] Reiter, R. (1968). Scheduling parallel computations. *Journal of the Association for Computing Machinery*, 15, No. 4:590–599.

[16] Schwiegelshohn, U. and Thiele, L. (1997a). Dynamic min-max problems. Technical Report 23, Swiss Federal Institute of Technology (ETH Zurich), TIK (Computer Engineering and Networks Laboratory). accepted for publication in Journal on Discrete Event Dynamic Systems.

[17] Schwiegelshohn, U. and Thiele, L. (1997b). Periodic and non-periodic mion-max equations. In *International Colloquium on Automata, Languages, and Programming*, ICALP 97, pages 379–389.

[18] Varshavsky, V., Kishinevsky, M., Kondratyev, A., Rosenblum, L., and Taubin, A. (1989). Models for specification and analysis of processes in asynchronous circuits. *Soviet Journal of Computer and System Sciences*, 26(2):61–76.

[19] Walkup, E. and Borriello, G. (1994). Interface timing verification with application to synthesis. In *IEEE/ACM Design Automation Conference*, pages 106–112.

[20] Yakovlev, A. (1992). On limitations and extensions of stg model for designing asynchronous control circuits. In *Proceedings of ICCD 92*, Cambridge, MA.

[21] Yakovlev, A., Kishinevsky, M., Kondratyev, A., Lavagno, L., and Pietkiewicz-Koutny, M. (1996). On the models for asynchronous circuit behaviour with OR causality. *Formal Methods in System Design*, 9:189–233.

[22] Yen, T., Ishii, A., Casavant, A., and Wolf, W. (1995). Efficient algorithms for interface timing verification. Technical report, Princeton University.

Chapter 5

LTRL-BASED MODEL CHECKING FOR A RESTRICTED CLASS OF SIGNAL TRANSITION GRAPHS

R. Meyer
LSV, ENS de Cachan
CNRS UMR 8643
61, avenue du Pdt Wilson
94235 Cachan Cedex
France
rmeyer@lsv.ens-cachan.fr

P.S. Thiagarajan* †
Chennai Mathematical Institute
92 G.N. Chetty Road
T. Nagar Chennai 600 017
India
pst@smi.ernet.in

Keywords: asynchronous circuits, linear time temporal logic, Petri nets, Signal Transition Graphs

1. INTRODUCTION

We show here that the linear time temporal logic LTrL [24] can be used as a specification mechanism for restricted signal transition nets[26]. The logic LTrL is interpreted over the finite prefixes of Mazurkiewicz traces.

* A part of this work has been supported by BRICS (Basic Research in Computer Science), CS Dept., Aarhus University, Aarhus,Denmark.
† A part of this work has been supported by the IFCPAR (Indo-French Centre for the promotion of Advanced Research) project 1502-1.

A. Yakovlev et al.(eds.), Hardware Design and Petri Nets, 93-106.
© 2000 *Kluwer Academic Publishers.*

As the reader may be aware,(Mazurkiewicz) traces are restricted labelled partial orders which constitute an elegant and powerful extension of the notion of a sequence. The theory of traces is rich and well-understood [3]. Traces provide the natural language for talking about the non-sequential linear time behaviour of 1-safe Petri nets, and various kinds of distributed transition systems [16]. Starting from [23] a variety of linear time temporal logics for traces have been investigated as surveyed in [8]. The main motivation for studying such logics is that they capture sequential properties that are "robust" w.r.t. interleavings (i.e. either all interleavings of a trace have the property or none do). It is known that such properties can be verified efficiently with the aid of partial order based reduction techniques of various kinds [17].

Among the logics for traces, LTrL occupies a special position in that it has exactly the same expressive power as the first order theory of traces [24]. Due to the classic theorem of Kamp [9], later strengthened by gabbay et.al. [6], this matches the expressive power of LTL, the linear time temporal logic formulated by Pnueli [18], in a sequential setting.

A natural application of LTrL would be to specify the non-interleaved linear time properties of systems modeled by, say, 1-safe Petri nets. Unfortunately, it turns out that the resulting model checking problem can be non-elementary hard. This follows easily from the non-elementary time lower bound established by Walukiewicz for the LTrL-satisfiability problem [25]. Here we show that LTrL could serve as a tractable specification language provided we restrict the system models against which we do model checking. One such restricted class of system models is suggested by the literature on Petri net models of asynchronous circuits. It turns out that if we consider so called signal transition nets whose underlying Petri nets are marked graphs [2] then the model checking problem can be solved in time $O(n^2 \times k^2 \times m)$. Here n is the size of the *state space* of the signal transition net (which will be in general exponential in the size of its presentation), k is the number of transitions in the underlying marked graph and m is the size of the specification formula.

Our main aim here is to establish a link between partial order based specification logics and net descriptions of asynchronous hardware. In this first attempt we have chosen LTrL as the specification logic mainly because, in terms of expressive power, it is a canonical logic. From a pragmatic standpoint it might be more fruitful to use the product version of LTL (see [8]) which has a clean theory, a model checking procedure which has the same complexity as LTL and is readily amenable to partial order reduction methods. As for the choice of the net model, we have been mainly motivated by the desire to work with a finite prefix of the unfolding of the net model while doing model checking. The simplest starting point is then obviously signal transition nets whose underlying Petri nets are marked graphs. A major disadvantage of course is that one can model only closed systems in this framework and thus can not handle inputs. As we point however in the concluding section, we believe that the methods proposed here can be extended to signal transition nets based on the restricted class of free choice nets known as bipolar synchronization schemes [7]. In this extended class, one can model inputs at least in a limited fashion. It is unlikely that our methods will scale up to signal transition nets based on arbitrary live and safe free choice nets; one should, in this case, choose a less expressive logic.

In terms of related work there is of course an extensive body of available work which study asynchronous hardware modeled by Petri nets. As for hardware models based on marked graphs one can go all the way back to M-nets of Seitz [22] and the later work on signal graphs by Rosenblum and Yakovlev[21]. There has also been a good deal of work on using unfoldings of net descriptions of asynchronous hardware to verify behavioural properties starting from McMillan [15] and with [10, 14, 12, 13, 11] containing a wealth of information on this topic. For an exhaustive account of the general literature we refer the reader to the excellent survey [26]. In terms of logical specifications of Petri net behaviour, the work of Esparza [4] and Reisig [19] are both relevant here. The logic considered by Esparza is a branching time one and it does not involve an until operator. In the work of Reisig, the logical concerns have more to do with coming up with a toolkit of sound proof rules for a logic which is a mix of temporal logic and UNITY logic [1]. In contrast what we do here - in a trace setting - is traditional linear time model checking.

In the next section we present the syntax and semantics of LTrL. In section 3 we identify the restricted class of signal transition nets: the underlying nets correspond to marked graphs. We also display some properties of such signal transition graphs expressible in LTrL. In section 4 we formulate the model checking problem and sketch the main ideas underlying our procedure. In the final section we discuss possible extensions of the preliminary work reported here.

2. THE LINEAR TIME TEMPORAL LOGIC LTRL

LTrL is a linear time temporal logic whose formulas are interpreted over the configurations of a trace. We recall briefly that a trace alphabet is a pair (Σ, D) where Σ is a finite non-empty set of actions and $D \subseteq$

$\Sigma \times \Sigma$ is a symmetric and reflexive dependency relation. Through the rest of this section we fix such a trace alphabet.

A trace over (Σ, D) is a Σ-labelled poset in which each element (viewed as an event) has a finite past and whose labeling function respects the dependency relation in a strong sense.

A Σ-labelled poset is a structure $(E, \leq, \lambda)$ where $(E, \leq)$ is a poset and $\lambda : E \to \Sigma$ is a labeling function. For $X \subseteq E$ we define $\downarrow X = \{y \mid \exists x \in X, y \leq x\}$ and write $\downarrow e$ instead of $\downarrow \{e\}$ for $e \in E$. The derived covering relation $\lessdot \subseteq E \times E$ is given by: $e \lessdot e'$ iff $e < e'$ and for every $e'' \in E$, if $e \leq e'' \leq e'$ then $e = e''$ or $e'' = e'$.

A trace over (Σ, D) is a Σ-labelled poset $\mathrm{Tr} = (E, \leq, \lambda)$ which satisfies, for every $e, e' \in E$, the following conditions:

(TR1) $\downarrow e$ is a finite set.

(TR2) $\lambda(e) D \lambda(e')$ implies $e \leq e'$ or $e' \leq e$.

(TR3) If $e \lessdot e'$ then $\lambda(e) D \lambda(e')$.

Let $\mathrm{Tr} = (E, \leq, \lambda)$ be a trace over (Σ, D). It is easy to see that E must be a countable set. Tr is said to be finite (infinite) in case E is a finite (infinite) set. A configuration of Tr is a subset c of E such that $\downarrow c = c$. Clearly, $\emptyset$, the empty set, is the least configuration (under $\subseteq$) of every trace. The configuration c is said to be finite in case c is a finite set. We let C_{Tr} denote the set of finite configurations of Tr. A natural transition relation one can associate with Tr is $\longrightarrow_{\mathrm{Tr}} \subseteq C_{\mathrm{Tr}} \times \Sigma \times C_{\mathrm{Tr}}$ defined via:

$$c \xrightarrow{a}_{\mathrm{Tr}} c' \text{ iff there exists } e \notin c \text{ such that } \lambda(e) = a \text{ and } c \cup \{e\} = c'.$$

It is easy to verify that if $c \xrightarrow{a}_{\mathrm{Tr}} c'$ and $c \xrightarrow{a}_{\mathrm{Tr}} c''$ then $c' = c''$.

The logic LTrL is parametrized by trace alphabets. In particular, the syntax of $\mathrm{LTrL}(\Sigma, D)$ is given by (while letting a range over Σ):

$$\mathrm{LTrL}(\Sigma, D) ::= p_a \mid \sim \alpha \mid \alpha \vee \beta \mid \langle a \rangle \alpha \mid \alpha \, \mathcal{U} \, \beta.$$

For convenience, we will often drop the mention of (Σ, D) in what follows. Let Tr be a trace and $c \in C_{\mathrm{Tr}}$ and α a formula of LTrL. Then the semantics of LTrL is specified via the inductive definition of when α is satisfied at c in Tr, denoted $\mathrm{Tr}, c \models \alpha$.

- $\mathrm{Tr}, c \models p_a$ iff there exists $e \in \max(c)$ such that $\lambda(e) = a$.

  Here $\max(c)$ is the set of maximal elements of c under $\leq$. Clearly $\max(c) = \emptyset$ iff $c = \emptyset$.

- $\mathrm{Tr}, c \models \sim \alpha$ and $\mathrm{Tr}, c \models \alpha \vee \beta$ are defined in the obvious way.

- $\text{Tr}, c \models \langle a \rangle \alpha$ iff there exists $c' \in C_{\text{Tr}}$ such that $c \xrightarrow{a}_{\text{Tr}} c'$ and $\text{Tr}, c' \models \alpha$.

- $\text{Tr}, c \models \alpha \, \mathcal{U} \, \beta$ iff there exists $c' \in C_{\text{Tr}}$ such that $c \subseteq c'$ and $\text{Tr}, c' \models \beta$. Further, if $c'' \in C_{\text{Tr}}$ such that $c \subseteq c'' \subset c'$ then $\text{Tr}, c'' \models \alpha$.

A formula α is satisfiable iff there exists a trace Tr and $c \in C_{\text{Tr}}$ such that $\text{Tr}, c \models \alpha$. We note that LTrL is essentially LTL, the usual linear time temporal logic of sequences if we set $D = \Sigma \times \Sigma$. In the next section we will consider examples of properties expressible in LTrL.

3.   MARKED SIGNAL TRANSITION GRAPHS

For the sake of brevity, we shall assume here familiarity with the basic notions and notations of net theory, including the notion of the unfolding of a net system. We shall lean on [20] as the basic reference for the notions we use here.

Our starting point will be $\mathcal{T}$-systems of the form $N = (P, T, F, M_{in})$ where (P, T, F) is a net satisfying $|{}^\bullet p| = |p^\bullet| = 1$ for every place $p \in P$. As usual M_{in} is the initial marking and given our purposes, it is convenient to assume that $M_{in} : P \to \{0, 1\}$. It should be clear that $\mathcal{T}$-systems are just net-based versions of marked graphs .

A marked signal transition graph (MSG) is a structure $G = (N, Y, \varphi, \mathcal{V}_0)$ where $N = (P, T, F, M_{in})$ is a $\mathcal{T}$-system which is required to be finite, strongly connected, live and safe. Y is a finite set of boolean signals and $\varphi : T \to Y \times \{+, -\}$ is a function which assigns a signal transition to each transition in T, and $\mathcal{V}_0 : Y \to \{0, 1\}$ is the initial valuation. Unlike the model suggested in [26], we require here N to be live and strongly connected only for the sake of convenience. A similar motivation holds for requiring φ to be total and limiting the range of φ to be $\{+, -\}$ instead of $\{+, -, \sim\}$ which would cater for toggle transitions (see [26] for related details). Through the rest of the paper we let $\star$ range over $\{+, -\}$ and write $y^\star$ instead of $(y, \star)$. In Fig. 5.1 we show a simple example of an MSG with $Y = (y_1, y_2)$ and $\mathcal{V}_0(y_1) = 0$ and $\mathcal{V}_0(y_2) = 1$.

We now wish to define - for want of a better term - the trace unfolding of an MSG. To this end, let $G = (N, Y, \varphi, \mathcal{V}_0)$ be an MSG with $N = (P, T, F, M_{in})$. Consider first the relation $D_N \subseteq T \times T$ defined as: $t_1 D_N t_2$ iff $({}^\bullet t_1 \cup t_1^\bullet) \cap ({}^\bullet t_2 \cup t_2^\bullet) \neq \emptyset$. Clearly (T, D_N) is a trace alphabet. Now suppose we take the unfolding of N and erase all occurrences of places. Then it is folklore that the resulting T-labelled prime event structure

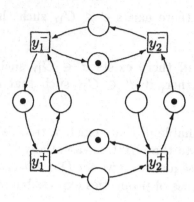

Figure 5.1   An simple examples of MSG

which will be conflict-free is an infinite trace over (T, D_N). We let TR_G denote this trace and call it the trace unfolding of G. For instance, the trace unfolding of the MSG shown on Fig. 5.1 will look as shown on Fig. 5.2.

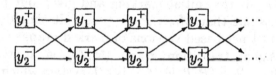

Figure 5.2   The trace unfolding of the MSG

We now use formulas taken from $\mathrm{LTrL}(T, D_N)$ to specify the dynamic properties of $G = (N, Y, \varphi, \mathcal{V}_0)$ with $N = (P, T, F, M_{in})$. We start with some derived modalities. Let $T_{y^\star} = \{t \mid \varphi(t) = y^\star\}$. Then $p_{y^\star} \overset{\Delta}{\Longleftrightarrow} \bigvee_{t \in T_{y^\star}} p_t$ and $\langle y^\star \rangle \alpha \overset{\Delta}{\Longleftrightarrow} \bigvee_{t \in T_{y^\star}} \langle t \rangle \alpha$. As usual $\Diamond \alpha \overset{\Delta}{\Longleftrightarrow} \mathsf{T} \, \mathcal{U} \, \alpha$ where $\mathsf{T} = p_{t_0} \vee \sim p_{t_0}$ for a fixed $t_0 \in T$. Further, $\Box \alpha \overset{\Delta}{\Longleftrightarrow} \sim \Diamond \sim \alpha$. Let $T_y = T_{y+} \cup T_{y-}$ for $y \in Y$ and consider the formula VLD^y specified as follows:

$$\mathrm{VLD}^y = \bigwedge_{\substack{t, t' \in T_y \\ t \neq t'}} \Box \sim (p_t \wedge p_{t'}).$$

In essence, this formula says that no two signal transitions associated with the signal y can ever occur concurrently in G. Next consider the formula VLD_{y^+} given by:

$$\text{VLD}_{y^+} = \sim \Diamond \langle y^+ \rangle (\sim p_{y^-} \, \mathcal{U} \, \langle y^+ \rangle \mathsf{T}).$$

This formula in essence says that it is not possible for a y^+-transition to be followed by another y^+-transition without an intervening y^--transition (strictly speaking, this interpretation holds only in the presence of the previously defined property VLD^y). We define VLD_{y^-} in a similar fashion. Finally, let INIT^y be the formula given by:

$$\text{INIT}^y = \begin{cases} (\sim p_{y^-}) \, \mathcal{U} \, \langle y^+ \rangle \mathsf{T} & \text{if } \mathcal{V}_0(y) = 0 \\ (\sim p_{y^+}) \, \mathcal{U} \, \langle y^- \rangle \mathsf{T} & \text{otherwise.} \end{cases}$$

This formula says that the first signal transition involving y must be y^+ (y^-) in case $\mathcal{V}_0(y) = 0$ ($\mathcal{V}_0(y) = 1$). Now let

$$\text{VLD} = \bigwedge_{y \in Y} (\text{VLD}^y \wedge \text{VLD}_{y^+} \wedge \text{VLD}_{y^-} \wedge \text{INIT}^y).$$

Then VLD says that every firing sequence of G starting from M_{in} is valid in the sense of [26]. In addition, one can also define a variety of (partial order based) liveness, safety and response properties using the operators $\Diamond$, $\Box$ and $\mathcal{U}$.

4.   THE MAIN RESULT

Let $G = (N, Y, \varphi, \mathcal{V}_0)$ be an MSG with $N = (P, T, F, M_{IN})$ and let α be a formula in $\text{LTrL}(T, D_N)$. Then we say that G meets the specification α - denoted $G \models \alpha$ - iff $\text{TR}_G, \emptyset \models \alpha$. The LTrL based model checking problem for MSGs is to determine, given such a pair G and α, whether or not $G \models \alpha$.

Theorem 1 *Let G and α be as above. Then the question $G \overset{?}{\models} \alpha$ can be settled in time $O(|T|^2 \cdot |\text{RM}_N|^2 \cdot |\alpha|)$ where $|T|$ is the number of transitions of N, $|\text{RM}_N|$ is the number of reachable markings of N and $|\alpha|$ is the size of the formula α.*

Since TR_G consists of a *single* infinite trace we can hope to develop a technique similar to the one employed in CTL-model checking.

Fix $G = (N, Y, \varphi, \mathcal{V}_0)$ with $N = (P, T, F, M_{in})$ and, as above, let RM_N be the set of reachable markings of N. For convenience we denote the set of finite configurations of $\text{TR}_G = (E, \leq, \lambda)$ by C_G. Clearly, there is a well-defined map mark : $C_G \to \text{RM}_N$ with $\text{mark}(\emptyset) = M_{in}$ and with the property $c \overset{t}{\longrightarrow} c'$ in TR_G iff $\text{mark}(c) \overset{t}{\longrightarrow} \text{mark}(c')$ in N.

Lemma 2 *There is an effectively computable positive constant $d \leq |T|$ and $E_1, E_2 \subseteq E$ (recall that $\mathrm{TR}_G = (E, \leq, \lambda)$) such that the following conditions hold:*

(i) *$E_1 \subset E_2$ and $E_1, E_2 \in C_G$. Further, $|C_1| \leq d.|\mathrm{RM}_N|$ and $|C_2| \leq 2d.|\mathrm{RM}_N|$ where $C_1 = \{c \in C_G \mid c \subseteq E_1\}$ and $C_2 = \{c \in C_G \mid E_1 \subseteq c \subseteq E_2\}$. Finally, $\mathrm{mark}(E_1) = \mathrm{mark}(E_2) = M_{in}$.*

(ii) *If $M \in \mathrm{RM}_N$ then there exists $c_1 \in C_1$ and $c_2 \in C_2$ such that $\mathrm{mark}(c_1) = M = \mathrm{mark}(c_2)$.*

(iii) *If $M, M' \in \mathrm{RM}_N$ such that M' is reachable from M then there exists $c \in C_1$ and $c' \in C_2$ such that $\mathrm{mark}(c) = M$ and $\mathrm{mark}(c') = M'$.*

This lemma follows from the standard theory of live and safe marked graphs. The bound $d \leq |T|$ is a conservative one. The exact required value of d will be determined by taking two successive copies of the finite prefix of TR_G as computed by, say, the procedure developed in [5]. We set E_2 to be the set of events of this "2-fold" prefix, say, $\mathrm{TR}_{\mathrm{fin}}$ and proceed as follows.

We will inductively label the members of $C_1 \cup C_2$ with the subformulas of α_0, the specification. We declare that $G \models \alpha_0$ iff at the end the empty configuration $\emptyset$ gets labelled with α_0.

We deal with subformulas of the form p_t in the obvious way. The cases $\sim\alpha$, $\alpha \vee \beta$ and $\langle a \rangle \alpha$ are also handled in the expected manner. The crunch case is when the subformula is of the form $\alpha\, \mathcal{U}\, \beta$.

First note that we can code every event e in the finite prefix by a unique pair (t, n), where $t = \lambda(e)$ is the label of e and $n = |\downarrow(e)|_t$ is the number of occurrences of t in the past of e (clearly $n \leq 2d$). Every configuration can be described by the set of its maximal events and hence we can code each configuration c by a unique and bounded set $s(c)$ such that

$$s(c) = \{(t, n) \mid (t, n) \text{ codes an event } e \text{ maximal in } c\}.$$

Thus every configuration in the prefix can be uniquely and finitely coded. Furthermore, it is easy to compute n for a fixed event e since the occurrences of t are totally ordered.

In the finite unfolding, we will start with the largest configurations and work our way backwards, level by level, where the level of a configuration is a measure of its size. In other words, we will dispose of all configurations of size m before considering configurations of size $m - 1$. Our algorithm will use a global set COL to collect all configurations where β is satisfied. Furthermore, for each configuration c the set $col(c)$

will represent all configurations of COL reachable from c. Finally, for each x in $col(c)$ we will compute $status(c, x)$ which says whether $\alpha \, \mathcal{U} \, \beta$ holds in c with x as a witness (i.e. α holds between c and x and β holds at x).

Now, the algorithm proceeds as follows:

1. Set $COL = \emptyset$ and $LEVEL = 0$ (0 represents for convenience the maximum level). Note that there will be a single configuration at level 0 and that the corresponding marking will be the initial one.

2. For each configuration c at the current level, do the following:

   (a) For every x in COL such that c has a successor c' satisfying $x \in col(c')$, x gets added to $col(c)$, which is initially empty.

   (b) For every x in $col(c)$, if one the following conditions is satisfied

       - c does not satisfy α.
       - c has a successor c' such that $x \in col(c')$ and $status(c', x) = 0$.

       then set $status(c, x) = 0$. Otherwise set $status(c, x) = 1$.

   (c) If c satisfies β, let $s(c)$ be the set associated with c. Then $s(c)$ is added to COL and to $col(c)$. Finally, set $status(c, s(c)) = 1$.

   (d) If there is an x in $col(c)$ such that $status(c, x) = 1$ then declare c to be good.

3. Move down to the next level.

At the end of this procedure we label all the good nodes with the until formula. Furthermore, if $c_1 \subseteq E_1$ and $c_2 \subseteq E_2$ are such that $mark(c_1) = mark(c_2)$ and c_1 has been labelled with the until fromula then we label c_2 also with the until formula.

Note that since the algorithm is working backwards, if two configurations c and c' are such that $c \subset c'$ then c' will be visited (at step 2) before c, since the level of c' is greater than the level of c. As a consequence, when a configuration c is being visited at step 2, all the configurations in the future of c have already been visited.

Now, step 2(c) ensures that every configuration satisfying β will eventually be added to the set COL (actually it is the encoding $s(c)$ of c that is added to COL). Thus, when a configuration c is being examined, every configuration greater that c satisfying β is already encoded in COL.

Step 2(a) ensures that every future configuration satisfying β and reachable from c is being encoded in $col(c)$. The special case where c itself satisfies β is dealt with in step 2(c).

Then step 2(b) ensures that $status(c, x) = 1$ if and only if the following conditions are met:

- c satisfies α.

- for every successor c' of c, if $x \in col(c')$ then $status(c', x) = 1$.

Thus only configurations x greater than c, satisfying β and such that α is satisfied all the way between c and x have $status(c, x)$ set to 1.

Finally step 2(d) declares c good iff there is a witness for c in $col(c)$, that is a configuration x such that $status(c, x) = 1$.

Now, since there are at most $2d|\mathrm{RM}_N|$ configurations in the finite prefix C_2, the algorithm will go at most $2d|\mathrm{RM}_N|$ times through step 2(a) to 2(d). However, each of those steps may require $O(2d|\mathrm{RM}_N|)$ time units to be performed, since the sets COL and $col(c)$ can contain at most $2d|\mathrm{RM}_N|$ elements. The whole labeling process for an until formula (provided the labeling is already done for the subformulas) will thus run in time $O(d^2|\mathrm{RM}_N|^2)$, that is, since $d \leq |T|$, in time $O(|T|^2|\mathrm{RM}_N|^2)$. It should be clear that the until formulas are the most time consuming formulas as far as labeling is concerned. Thus, if $|\alpha|$ is the size of the formula that we want to model-check, the whole process will run in time $O(|T|^2|\mathrm{RM}_N|^2|\alpha|)$.

The average space complexity can be improved by working directly with the state space of N and computing $\mathrm{TR}_{\mathrm{fin}}$ "backwards" on-the-fly for each subformula of the form $\alpha \, \mathcal{U} \, \beta$. A considerable simplification can also achieved for specifications which do not involve the use of the until operator and instead involve only the derived $\Diamond$ and $\Box$ modalities.

5.   CONCLUSION

As pointed out in the introduction, the circuit model we consider is very restricted in the sense that it does not model inputs. An obvious way out is to consider signal transition nets based on free choice nets [7]. In this case however the linear time behaviour of the system will consist of more than one (and in general infinitely many) maximal traces.. Hence the state labeling technique we have used here will not work. It is also quite likely that in this case the model checking problem based on LTrL has high complexity.

One can however consider the subclass of free choice nets called bipolar synchronization schemes (bp schemes). These restricted free choice nets are well structured in the sense that choice and synchronization must be well nested [7]. In Fig. 5.3we show an example of a bp scheme. As

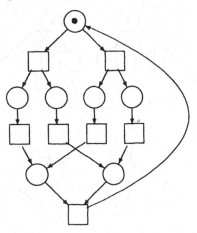

Figure 5.3 Example of a bp scheme

this example shows, the "AND-OR" nesting we have in mind can be somewhat subtle. Clearly these nets can model inputs. The key point about bp schemes is that their behaviour can be symbolically represented by a live and safe marked graph as shown in [7]. Consequently we expect a more involved version of the procedure presented here to go through for signal transition nets based on bp schemes. What is not clear is how rich is this class of signal transition nets. In Fig. 5.4 we show a typical example of a free choice net which is *not* a bp scheme. This example seems to suggest that independently arriving inputs can not interact freely. An interesting question is whether a natural restriction on input-output modes can be used to characterize the class of signal transition nets based on bp schemes. This will let us then judge how useful are this class of circuit models.

References

[1] K. Mani Chandy and Jayadev Misra. *Parallel Program Design : a Foundation.* Addison-Wesley, Reading, Mass., 1988.

[2] F. Commoner, A. W. Holt, S. Even, and A. Pnueli. Marked directed graphs. *J. Comput. System Sci.*, 5(1):511–523, October 1971.

[3] V. Diekert and G. Rozenberg. *The Book of Traces.* World Scientific, Singapore, 1995.

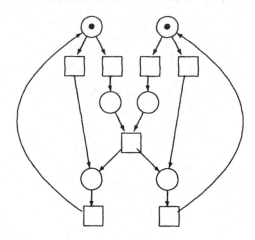

Figure 5.4   Example of a fc net which is not a bp scheme

[4] J. Esparza. Model checking using net unfoldings. *Science of Computer Programming*, 23(2–3):151–195, December 1994.

[5] J. Esparza, S. Römer, and W. Vogler. An improvement of McMillan's unfolding algorithm. In *Proc. TACAS '96*, volume 1055 of *Lecture Notes in Computer Science*, pages 87–106. Springer-Verlag, 1997.

[6] D. Gabbay, A. Pnueli, S. Shelah, and J. Stavi. On the temporal basis of fairness. In *Conference Record of the Seventh Annual ACM Symposium on Principles of Programming Languages*, pages 163–173, Las Vegas, Nevada, January 1980.

[7] H.J. Genrich and P.S. Thiagarajan. A theory of bipolar synchronization schemes. *Theoretical Computer Sience*, 30:241–318, 1984.

[8] J. G. Henriksen and P. S. Thiagarajan. Distributed versions of linear time temporal logic: A trace perspective. In *Lectures on Petri Nets I*, volume 1491 of *Lecture Notes in Computer Science*, pages 643–682. Springer-Verlag, 1998.

[9] H. Kamp. *On tense logic and the theory of order*. PhD thesis, University of California, 1968.

[10] M. Kishinevsky, A. Kondratyev, A. Taubin, and V. Varshavsky. *Concurrent Hardware: The Theory and Practice of Self-Timed Design*. Series in Parallel Computing. John Wiley & Sons, 1994.

[11] A. Kondratyev, J. Cortadella, M. Kishinevsky, L. Lavagno, A. Taubin, and A. Yakovlev. Identifying state coding conflicts in asynchronous system specifications using petri net unfoldings. In *International Conference on Application of Concurrency to System Design*, pages 152–163, Japan, March 1998.

[12] A. Kondratyev, M. Kishinevsky, A. Taubin, and S. Ten. A Structural Approach for the Analysis of Petri Nets by Reduced Unfoldings. In *17th International Conference on Application and Theory of Petri Nets*, pages 346–365, Osaka, Japan, June 1996. Springer-Verlag.

[13] A. Kondratyev, M. Kishinevsky, A. Taubin, and S. Ten. Analysis of Petri nets by ordering relations in reduced unfoldings. *Formal Methods in System Design*, 12(1):5–38, 1998.

[14] A. Kondratyev and A. Taubin. On verification of the speed-independent circuits by STG unfoldings. In *International Symposium on Advanced Research in Asynchronous Circuits and Systems*, Salt Lake City, Utah, USA, November 1994.

[15] K.L. McMillan. Using unfoldings to avoid the state explosion problem in the verification of asynchronous circuits. In *Proc. CAV '92, Fourth Workshop on Computer-Aided Verification*, volume 663 of *Lecture Notes in Computer Science*, pages 164–174. Springer-Verlag, 1992.

[16] M. Nielsen and G. Winskel. Trace structures and other models for concurrency. In V. Diekert and G. Rozenberg, editors, *The Book of Traces*, chapter 9, pages 271–305. World Scientific, Singapore, 1995.

[17] D. Peled. Partial order reduction: Model-checking using representatives. In Wojciech Penczek and Andrzej Szalas, editors, *Mathematical Foundations of Computer Science 1996*, volume 1113 of *Lecture Notes in Computer Science*, pages 93–112. Springer-Verlag, 1996.

[18] A. Pnueli. The temporal logic of programs. In *Proceedings of the 18th IEEE Symposium on the Foundations of Computer Science (FOCS-77)*, pages 46–57, Providence, Rhode Island, October 31–November 2 1977. IEEE, IEEE Computer Society Press.

[19] W. Reisig. *Elements of Distributed Algorithms*. Springer-Verlag, Berlin, 1998.

[20] W. Reisig and G. Rozenberg, editors. *Lectures on Petri nets, Parts I and II*, volume 1491 and 1492 of *Lecture Notes in Computer Science*. Springer-Verlag, 1998.

[21] L.Y. Rosenblum and A.V.Yakolev. Signal graphs : from self-timed to timed ones. In *Proc. of International workshop on Timed Petri nets*, pages 199–207. IEEE Computer Society Press, 1985.

[22] C.L. Seitz. Asynchronous machines exhibiting concurrency. In *Proc. of Project MAC conference on concurrent systems and parallel computation*, 1970.

[23] P. S. Thiagarajan. A trace based extension of linear time temporal logic. In *Proceedings, Ninth Annual IEEE Symposium on Logic in Computer Science*, pages 438–447, Paris, France, 4–7 July 1994. IEEE Computer Society Press.

[24] P. S. Thiagarajan and I. Walukiewicz. An expressively complete linear time temporal logic for Mazurkiewicz traces. In *Proceedings, Twelth Annual IEEE Symposium on Logic in Computer Science*, pages 183–194, Warsaw, Poland, 29 June–2 July 1997. IEEE Computer Society Press.

[25] I. Walukiewicz. Difficult configurations — on the complexity of LTrL. In *Proceedings of ICALP'98*, volume 1443 of *Lecture Notes in Computer Science*. Springer-Verlag, 1998.

[26] A. V. Yakovlev and A. M. Koelmans. Petri nets and digital hardware design. In *Lectures on Petri Nets I*, volume 1491 of *Lecture Notes in Computer Science*, pages 154–236. Springer-Verlag, 1998.

Chapter 6

A POLYNOMIAL ALGORITHM TO COMPUTE THE CONCURRENCY RELATION OF A REGULAR STG

Andrei Kovalyov
Department of Electrical and Computer Engineering
University of Manitoba
Winnipeg, Manitoba R3T 2N2
Canada
kovalyov@ee.umanitoba.ca

Abstract   Some problems of analysis and synthesis of signal transition graphs and Petri nets are studied in this article. Several synthesis techniques for STGs have been proposed which require knowledge of the concurrency relation of the net, i.e., the pairs of transitions that can become concurrently enabled at some reachable marking. With the algorithm in [KE] we can compute the concurrency relation on free choice signal transition graphs [C]. In this article we generalize this method to signal transition graphs, which are not necessarily free-choice, as for example, [LS,M]. The proof is closely related to a reduction method. We generalize and improve a reduction method given in [E1] in three more aspects.

Keywords: Asynchronous circuits, Petri nets, Signal Transition Graphs

1.   INTRODUCTION

Signal Transition Graphs (STGs) have become a popular and much studied formalism for the modeling, specification and verification of speed independent circuits [C,LS,M,Ya]. STGs are bounded Petri nets whose transitions are labeled with the changes of binary signals. The occurrence of a transition with label y^+ *raises* y, while the occurrence of a transition with label y^- *lowers* y.

To produce hazard-free implementations, STGs must be *consistently encoded* [C].

A. Yakovlev et al.(eds.), Hardware Design and Petri Nets, 107-126.
© *2000 Kluwer Academic Publishers.*

In current synthesis tools, the verification of these properties is carried out by constructing the reachability graph of the STG and therefore is very expensive computationally.

Recently, some synthesis methods have been proposed for Free-Choice STGs which avoid this construction [PC, KPCR]. They have low polynomial complexity, but require knowledge of which pairs of transitions of the net can become concurrently enabled. Therefore, an efficient polynomial algorithm for the computation of the concurrency relation between transitions has become very important for these methods to be applicable. In general to find such an algorithm and to prove it for nontrivial classes of Petri nets is extremely hard task. The problem is EXPSPACE-hard for arbitrary systems, and PSPACE-complete for 1-bounded systems [CEP] (we use the terms Petri net and system interchangeably).

In [KE] we propose an $O(n^3)$ algorithm for the computation of the concurrency relation of extended Free-Choice Petri nets. Before that time the problem has been shown to be polynomial only for live Marked Graphs [K1] and 1-bounded Conflict-Free systems [Ye,E2] (the algorithm of [E2] can be easily generalized to the n-bounded case).

The article [Ya] contains a strong argument in favor of lifting Free-Choice limitation for Signal Transition Graphs.

In this article we generalize the algorithm for the computation of the concurrency relation to Regular Petri nets, and therefore to Regular STGs. The time complexity of the algorithm is $O(n^4)$, where n is the number of nodes of the net.

Regular Petri nets are those systems that satisfy the conditions of the Rank Theorem [D1], which refers to the linear algebraic representation of systems. Regularity is a sufficient condition for an ordinary system to be live and bounded [D1]. Notice that Regularity of Petri nets can be efficiently decided (see, for instance, [K2]).

While being wider than the class of live and bounded Free-Choice systems, the class of Regular systems is still restricted. The usefulness of Regular systems in the design flow can be shown as follows. We propose to identify the correctness property of a circuit as Regularity of the corresponding Petri net. Then we can use modular design of a correct model similar to the one for Free-Choice Petri nets [ES].

Regular systems are a subclass of a wider class of State Machine Decomposable systems (SMD-systems). An SMD-system consists of several state machines. In practice, the design of systems traditionally uses state machine model which is very familiar to engineers and designers. Often the system consists of several concurrently running and communicating state machines. Each subsystem (state machine) can be represented in

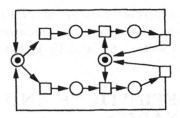

Figure 6.1 Smaller Version of VME-bus controller from [Ya].

the model autonomously. Then such representations should be coordinated to describe the whole system. It is modular approach to system design. There is a broad spectrum of models supporting modular approach for system specification, for instance, SDL [FO], CSP [Ho], CCS [Mi], LCS [LW], COSY [LTS].

The usefulness of our algorithm is raised by conjecture that it is valid for the whole class of SMD-systems. The formal proof is left for future work. For instance, the Signal Transition Graph from [Ya] (smaller version of it is shown on Fig. 6.1), which models the VME-bus controller is SMD, and our algorithm is valid for it.

For the proof of the concurrency relation method we use reduction. In this article we improve the complete reduction method of [E1]. We develop the method in three more aspects. First, we apply the method for a wider class of systems. We give a reduction rule and prove its strong soundness and completeness for Regular systems. Second, we give more precise evaluation of complexity of this method ($O(|S| \times |T|)$), comparing to [E1] (polynomial in [E1] without specifying the degree of polynomial). Third, in the method of [E1] one of the reduction rules ($R3$) checks if all the siphons are nonempty at every step of its application in order to be strongly sound (original system is live and bounded iff reduced system is live and bounded). This check is cumbersome and actually is necessary only once at the first application of $R3$ because at each step of the reduction all the deadlocks are marked. In our method such a check is not required.

To simplify the reduction process and the corresponding propositions, we will use only one reduction rule R instead of two or more as in [DE,E1]. In this case we are forced to use live and bounded Marked Graphs as atomic systems instead of strongly connected system consisting of two nodes. Analysis of Marked Graphs is very simple [AFN].

After some basic definitions and results (Section 2), we give a reduction rule and prove its soundness for T-coverable systems, and its completeness for Regular systems (Section 3). Then we provide an algorithm of reduction and find its complexity (Section 4). In Section 5, we define the structural concurrency relation of a Petri net. Then we prove that the concurrency and the structural concurrency relation coincide for Regular Petri nets (Section 6), and that they can be computed in $O(n^4)$ time (Section 7).

2.   BASIC DEFINITIONS AND PRELIMINARIES

A *net* is a triple $N = (S, T, F)$ with $S \cap T = \emptyset$ and $F \subseteq (S \times T) \cup (T \times S)$. S is the set of *places*, T is the set of *transitions*, $X = S \cup T$ is the set of nodes of N. For $x \in X$, the pre-set ${}^\bullet x$ is defined as ${}^\bullet x = \{y \in X | (y, x) \in F\}$ and the post-set $x^\bullet$ is defined as $x^\bullet = \{y \in X | (x, y) \in F\}$. For $Y \subseteq X$, ${}^\bullet Y = \cup_{x \in Y} {}^\bullet x$ and $Y^\bullet = \cup_{x \in Y} x^\bullet$.

From now on we assume that all nets we deal with are *finite* (i.e. X is finite), *connected* (i.e. $X \times X = (F \cup F^{-1})^*$, where R^* denotes the reflexive and transitive closure of a relation R), and have at least one place and one transition, with exception of CP-subnets defined below, which can consist of single transition. N is *strongly connected* iff $X \times X = F^*$.

A *marking* of a net $N = (S, T, F)$ is defined as a mapping $M : S \to \mathbb{N}$. A place is called *marked* by M iff $M(s) > 0$. Let $S' \subseteq S$. Then S' is *marked* iff $\exists s \in S'$ s is marked.

A *system* is a pair (N, M^0). On Fig. 6.1 a system is shown. Places are graphically represented by circles, transitions by boxes, arcs of the relation F by arrows and a marking by placing $M(s)$ tokens (black dots) into the corresponding place.

A transition $t \in T$ is *fireable* at a marking M (denoted $M[t\rangle$) iff $\forall s \in {}^\bullet t$ $M(s) > 0$. Firing of t yields a new marking M' (denoted $M[t\rangle M'$), where

$$M'(s) = \begin{cases} M(s) + 1 & \text{iff } s \in t^\bullet \backslash {}^\bullet t \\ M(s) - 1 & \text{iff } s \in {}^\bullet t \backslash t^\bullet \\ M(s), & \text{otherwise} \end{cases}$$

The empty sequence ϵ is an occurrence sequence: we have $M[\epsilon\rangle M$ for every marking M.

A transition t is *live* iff for every reachable marking M there exists a marking M' reachable from M which enables t. A marking M is called *live* iff every transition is live in (N, M). A place $s \in S$ is *n-bounded* ($n \in \mathbb{N}$) iff for every reachable marking M, $M(s) \leq n$; s is *bounded* iff $\exists n \in \mathbb{N}$ such that s is n-bounded. A marking M is *bounded* iff every place is bounded in (N, M). A net N is *well-formed* iff there exists a live and bounded marking of N.

Let S and T be arbitrarily but fixed ordered. Let $Q \subseteq S (Q \subseteq T)$. Then we denote $\chi(Q)$ the characteristic vector of the set Q. The incidence matrix $C : S \times T \rightarrow \{-1,0,1\}$ of N is defined by $C(-,t) = \chi[t^\bullet] - \chi[^\bullet t]$. We denote every vector $(0,0,...,0)$ by $\bar{0}$ and every vector $(1,1,...,1)$ by $\bar{1}$. A vector J is called *S-invariant(T-invariant)* iff $J \cdot C = \bar{0}$ $(C \cdot J = \bar{0})$. For two vectors J and J', $J \geq J'$ iff for every i-th component $J(i) \geq J'(i)$. An invariant J is *semi-positive* iff $J \geq 0$ and $J \neq \bar{0}$. An invariant J is *positive* iff $J \geq \bar{1}$. The *support* $\langle J \rangle$ of a vector J is the set of elements x satisfying $J(x) \neq 0$. For a set $Q \subseteq T$, $dim(Q)$ denotes the dimension of the space of such T-invariants J which satisfy $\forall t \notin Q \quad J(t) = 0$.

A *cluster* of a net N is a connected component of the relation $F_1 \cup F_1^{-1}$, where $F_1 = F \cap (S \times T)$. $[x]$ denotes the cluster containing $x \in S \cup T$). The set of all clusters is denoted by $\mathcal{A}$.

A net $N = (S, T, F)$ is called an *EFC-net* iff $\forall s \in S \; \forall t \in T \; (s, t) \in F$ implies $^\bullet t \times s^\bullet \subseteq F$. A system (N, M^0) (not necessarily EFC) is called *Regular* [D1] iff the following four conditions hold.

$$\text{there exists a positive S-invariant} \qquad (6.1)$$

$$\text{there exists a positive T-invariant} \qquad (6.2)$$

$$rank(C) = |\mathcal{A}| - 1 \qquad (6.3)$$

$$\text{for every semi-positive S-invariant } I \text{ of } N, I \cdot M^0 > 0 \qquad (6.4)$$

In [D2] a necessary and sufficient condition of liveness and boundedness of EFC-nets is given. It is the Rank Theorem.

Theorem 2.1 [D2] Let N be an EFC-net. Then (N, M^0) is live and bounded iff (N, M^0) is Regular. $\square$

The Rank Theorem does not hold for arbitrary nets, but provides a sufficient condition for liveness and boundedness.

Let $[x]$ be a cluster of N. A *feedback* [DE] of $[x]$ is an arc $(t, s) \in F$ such that $s \in [x], t \in [x], (s, t) \notin F$. A net N is *feedback-free* [DE] iff no cluster has feedbacks.

Lemma 2.2 [D1] Every Regular system is live, bounded and feedback-free. $\square$

Let N be feedback-free. Its *EFC-representation* [D1] N' is defined by $N' = (S, T, F \cup \widehat{F} \cup \widehat{F}^{-1})$ where $\widehat{F} = \{(s, t) \in S \times T \mid [s] = [t] \wedge (s, t) \notin F\}$ and $\widehat{F}^{-1} = \{(t, s) | (s, t) \in \widehat{F}\}$.

Lemma 2.3 [D1] Let N be feedback-free and N' its EFC-representation. Then

(a) N' is an EFC-net.

(b) $C = C'$, i.e. both nets have identical incidence matrices. $\square$

Now we give a necessary and sufficient condition of Regularity expressed in terms of the EFC-representation.

Lemma 2.4 [K2] Let N be a net and N' is the EFC-representation of N. Then (N, M) is Regular iff (N', M) is live and bounded and N is feedback-free. $\square$

A net (S, T, F) is a *T-net* iff it satisfies two following conditions

$$\forall s \in S \quad |s^\bullet \cap T| = 1 \tag{6.5}$$

$$\forall s \in S \quad |{}^\bullet s \cap T| = 1 \tag{6.6}$$

Let $N = (S, T, F)$, $S' \subseteq S$, $T' \subseteq T$. Then (S', T', F') is a *subnet* of N iff $F' = F \cap ((S' \times T') \cup (T' \times S'))$. Since F' is completely defined by S' and T', we will write (S', T') instead of (S', T', F'). A set $T_1 \subseteq T$ is a *T-component* iff $({}^\bullet T_1, T_1)$ is a strongly connected subnet of (S, T, F) satisfying 6.5, 6.6 and

$$T_1^\bullet = {}^\bullet T_1 \tag{6.7}$$

Let $N = (S, T, F)$ be a net. Then $N^{-d} = (T, S, F^{-1})$ is *reverse-dual* net of net N. The notion of an *S-component* is reverse-dual to that of a T-component. A net is *T-coverable* iff every transition of it belongs to some T-component. We call a cover by T-components by *T-cover*.

Lemma 2.5 Let N be a net.

(a) [DE] If T_1 is a T-component of N then $\chi(T_1)$ is a minimal T-invariant of N.

(b) [DE] If J is a minimal T-invariant of a well-formed EFC-net N then the support $\langle J \rangle$ is a T-component of N.

(c) [MR] If J is a semi-positive $S - (T-)$invariant then the support $\langle J \rangle$ satisfies the following condition: $^\bullet\langle J \rangle = \langle J \rangle^\bullet$. $\square$

Lemma 2.6 Let N be a feedback-free T-coverable net and not a T-net, and N' be its EFC-representation.

(a) $F \subseteq F'$.

(b) Nodes $dom(F' \backslash F) \cup cod(F' \backslash F)$ belong to clusters with more than one place and more than one transition in both N and N'.

(c) If a net N satisfies 6.1, 6.2, 6.3 then T_1 is a T-component of N iff T_1 is a T-component of N'.

Proof: (a) and (b) easy to follow from the definition of the EFC-representation.

(c) Let N satisfies 6.1, 6.2, 6.3. Then by Lemma 2.2, it is feedback-free.

$\Leftarrow$. Let T_1 be a T-component of N'. Then by Lemmata 2.3b and 2.6a, the subnet $(^\bullet T_1, T_1)$ of N has the same set of transitions T_1, but (possibly) does not have all the self-loops compared to the subnet $(^\bullet T_1, T_1)$ of N'. So the subnet $(^\bullet T_1, T_1)$ of N is strongly connected and satisfies 6.5 and 6.6. By Lemma 2.5a, $\chi(T_1)$ is a minimal T-invariant of N'. By Lemma 2.3b, the incidence matrices of N and N' are the same, and hence $\chi(T_1)$ is a minimal T-invariant of N. By Lemma 2.5c, $^\bullet T_1 = T_1^\bullet$ in N. Hence by definition, T_1 is a T-component of N.

$\Rightarrow$. Let T_1 be a T-component of N. By Lemma 2.5a, $\chi(T_1)$ is a minimal T-invariant of N. By Lemma 2.3b, the incidence matrices of N and N' are the same, and hence $\chi(T_1)$ is a minimal T-invariant of N'. By Lemma 2.5b, T_1 is a T-component of N'. $\square$

Let N be a net. The net $\phi(N)$ is defined as the result of performing the following operations for every cluster $[x]$ of N and every feedback (t, s) of $[x]$:

(i) remove the arc (t, s);

(ii) add a new place s_1 and a new transition t_1;

(iii) add arcs (t, s_1), (s_1, t_1) and (t_1, s).

It is easy to see that this transformation always terminates, and does not depend on the order in which the feedbacks are treated. The following lemma provides some properties of the transformation $\phi(N)$.

Lemma 2.7 Let N be a net.

(a) [DE] The net $\phi(N)$ is feedback-free.

(b) If N is T-coverable so is $\phi(N)$.

Proof: (b) is easy. $\square$

Lemma 2.8 Let N be a net.

(a) [B] If N is well-formed then it is strongly connected.

(b) [Ha] If an EFC-net is well-formed then it is T-coverable.

(c) [KL] If an EFC-net is strongly connected and $\exists J \geq 1 \quad C \cdot J \leq 0$ then $rank(C) \geq |\mathcal{A}| - 1$. $\Box$

Lemma 2.9 Let N be a net.

(a) If a net is feedback-free and T-coverable then $rank(C) \geq |\mathcal{A}| - 1$.

(b) If a net satisfies 6.1, 6.2, 6.3 then it is T-coverable.

(c) $rank(C) = |T| - dim(T)$.

(d) If a net N satisfies 6.1, 6.2, 6.3 then S_1 is an S-component of N iff S_1 is a minimal S-invariant of N.

Proof: (a) Let N' be the EFC-representation of N. By definition of the EFC-representation, the number of clusters in N and N' is the same. By Lemma 2.3b, rank of the incidence matrices of N and N' is the same. Since N is T-coverable, it is strongly connected and satisfies $\exists J \geq 1 \quad C \cdot J \leq 0$. The result follows from Lemma 2.8c.

(b) follows from Lemmata 2.8b and 2.6c.

(c) is well-known from Linear Algebra.

(d) follows from Lemmata 2.5ab, 2.6c and reverse-duality. $\Box$

Theorem 2.10 [KL] A net N satisfies 6.1, 6.2, 6.3 iff it is T-coverable and 6.3 holds. $\Box$

3.   A REDUCTION RULE, ITS STRONG SOUNDNESS FOR T-COVERABLE SYSTEMS AND ITS COMPLETENESS FOR REGULAR SYSTEMS

Let N be a T-coverable net. Then non-empty and connected subnet $\widehat{N} = (\widehat{S}, {}^\bullet\widehat{S} \cup \widehat{S}^\bullet)$ is a *CP-subnet* of N iff

(i) there exists a T-cover C of N and a T-component $N_1 \in C$ such that

(ii) $\widehat{N} \subset N_1$.

(iii) $\overline{N} = (\overline{S}, \overline{T}) = (S \backslash \widehat{S}, T \backslash ({}^\bullet\widehat{S} \cup \widehat{S}^\bullet))$ contains some transition, and is strongly connected.

We denote $\widehat{T}_i = \overline{S}^\bullet \cap ({}^\bullet\widehat{S} \cup \widehat{S}^\bullet)$, $\overline{\mathcal{A}}_o$ the set of clusters of $\overline{N}$ which have common nodes with $\widehat{N}$ in N, and $\widetilde{\mathcal{A}}$ the set of clusters of N which have common nodes in $\overline{N}$ and $\widehat{N}$.

Theorem 3.1 [DE] Let N be a well-formed EFC-net and $\widehat{N}$ be a CP-subnet of N. Then $(\overline{S}, \overline{T})$ is well-formed and

$$|\widehat{T}_i| = 1 \tag{6.8}$$

$\Box$

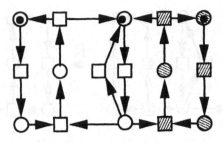

Figure 6.2 A Regular system and one of its CP-subnet (dashed nodes).

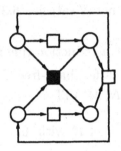

Figure 6.3 Violation of Condition 6.9.

The system on Fig. 6.2 is a Regular system (a model of non-deterministic wait process from [Mu]), and one of its CP-subnets (dashed nodes) satisfies 6.8.

For the class of T-coverable nets we may notice that some transitions $t_i \in \widehat{T}_i$ may have pre-sets ${}^\bullet t_i$ containing the places of different clusters of $\overline{N}$, because we no longer have the EFC-property. In other words, after removal of a CP-subnet, the number of clusters in the resulting net $\overline{N}$ may increase. On Fig. 6.3 we show the example, where dashed transition represents a CP-subnet and shares two its input places with two clusters of $\overline{N}$. So we can have the situation when $\overline{N}$ satisfies 6.3, but N does not or vice versa. To preserve Rank equation we need an additional condition:

$$|\overline{\mathcal{A}}_o| = 1 \qquad (6.9)$$

Lemma 3.2 Let $\widehat{N}$ be a CP-subnet of a net N. Then
   (a) $\widehat{N}$ is a T-net
   (b) $Rank(\widehat{C}) = |\widehat{T}| - 1$
   (c) $Rank(C) \geq Rank(\overline{C}) + Rank(\widehat{C})$
   (d) $dim(T) \geq dim(\overline{T}) + 1$

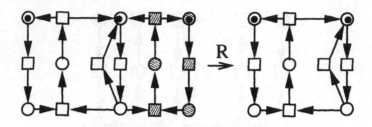

Figure 6.4   Reduction Rule R.

(e) If a CP-subnet $\widehat{N}$ of a net N satisfies 6.8 then $|\mathcal{A}| = |\overline{\mathcal{A}}| - |\overline{\mathcal{A}}_o| + |\widehat{T}|$

Proof: in [K3]. □

Now we are ready to give a reduction rule for T-coverable systems.

Rule *R.* Let (N, M^0) be a feedback-free T-coverable system and not a T-system. $(\overline{N}, \overline{M}^0) = R(N, M^0)$ iff
Conditions on (N, M^0).

C1. N has a CP-subnet $\widehat{N}$ of N with 6.8, 6.9 and

$$(\overline{S}, \overline{T}) \text{ is } T - coverable \qquad (6.10)$$

C2. Every loop of $\widehat{N}$ is marked.

Changes in (N, M^0) to produce $(\overline{N}, \overline{M}^0)$.

R1. $\forall s \in \widehat{T}^{\bullet} \cap \overline{S}$

$$\overline{M}^0(s) = \begin{cases} M^0(s) & \text{iff } \exists \text{ a non-marked path from } \widehat{T}_i \text{ to } s \text{ in } \widehat{N} \\ M^0(s) + 1, & \text{otherwise} \end{cases}$$

R2. Remove $\widehat{N}$ from N.

Fig. 6.4 illustrates Reduction Rule R.

Theorem 3.3 (strong soundness of R) R is strongly sound with respect to the class of Regular systems.

Proof: in [K3]. □

To simplify the reduction process and the corresponding propositions, we will use only one reduction rule instead of two or more as in [DE,E1]. In this case we are forced to use live and bounded T-systems as atomic nets instead of strongly connected system consisting of two nodes. Analysis of T-systems is very simple [AFN].

Theorem 3.4 (completeness of R) R is complete for the class of Regular systems.

Proof: in [K3]. □

4. THE ALGORITHM OF REDUCTION AND ITS COMPLEXITY

The algorithm below has two passes (two while loops). During the first pass, the algorithm checks if a given system (N, M^0) is T-coverable, finds a T-cover and corresponding CP-subnets, and simultaneously memorizes CP-subnets in a STACK. Actual reduction is done during the second pass (second while-loop).

Algorithm 1 (Algorithm of Reduction)
Input: (N, M^0) is a system.
Output: one of the following two messages: (1) "(N, M^0) is completely reduced " or (2) "(N, M^0) is NOT completely reduced ".
Function: *get-T-component(N, t)*. This function finds an T-component of N containing transition t. It is similar (and reverse-dual) to the function *mark-S-component* [Kem], but outputs the subnet $(\overline{S}, \overline{T})$ of transitions of a T-component containing t. If it finds a minimal set $T_1 \subseteq T$ such that $T_1^\bullet \subseteq {}^\bullet T_1$ and $T_1^\bullet \neq {}^\bullet T_1$, the algorithm stops and outputs a message "(N, M^0) is NOT completely reduced".
Function: *strongconnect(N)*. This function is based on the depth-first search algorithm [Ta]. It checks if the graph N is strongly connected.
Function: *get-CP-subnet$(\overline{N}, N_1, t)$*. This function finds a CP-subnet $\widehat{N} \subseteq N_1$ of $\overline{N} \cup N_1$ containing t.
begin
    if *strongconnect(N)* $= No$
        then Stop with "N is NOT completely reduced"
    endif
    choose $t \in T$;
    $(\overline{S}, \overline{T}) := get - T - component(N, t)$;
    $T^o := \overline{S}^\bullet \backslash \overline{T}$;
    while $(T^o \neq \emptyset)$ **do**
        choose $t \in T^o$; $T^o := T^o \backslash \{t\}$;
        $T_1 = get - T - component(N, t)$;
        $(\widehat{S}, \widehat{t}) := get - CP - subnet(\overline{N}, N_1, t)$;
        $\overline{N} := \overline{N} \cup \widehat{N}$; $Push(STACK, \widehat{N})$;
    endwhile
    while $(STACK \neq \emptyset)$ **do**
        $\widehat{N} := Pop(STACK)$;
        if $\widehat{N}$ and M^0 satisfy (6.8, 6.9) and C2
            then $(N, M^0) := R(N, M^0)$;
            else Stop with "(N, M^0) is NOT completely reduced"
        endif
    endwhile

 if (N, M^0) is live and bounded T-net
  then Stop with "(N, M^0) is completely reduced";
   else Stop with "(N, M^0) is NOT completely reduced";
  endif
end

Proposition 4.2 Algorithm 1 has the complexity $O(|S| \times |T|)$.

 Proof: The complexity of function "strongconnect" for a directed graph $G = (V, E)$ is $O(|V| + |E|)$ [Ta]. The complexity for function "mark-S-component" is given in [Kem] ($O(|T|)$). Since the set of the CP-subnets (under consideration) defines a partition on the set of nodes of the net, the complexity of the finding and checking of the CP-subnets is $O(|S| \times |T|)$. Hence the upper bound of the time complexity of the algorithm is $O(|S| \times |T|)$. The complexity of verifying if a given T-system is live and bounded is $O(|S| \times |T|)$. $\square$

5. CONCURRENCY RELATIONS

 The concurrency relation is usually defined as a set of pairs of transitions. We use a more general definition.

 Let (N, M_0) be a net system, and let X be the set of nodes of N. Given $x \in X$, define the marking M_x of N as follows:

- if x is a place, then M_x is the marking that puts one token on x, and no tokens elsewhere;

- if x is a transition, then M_x is the marking that puts one token on every input place of x, and no tokens elsewhere.

The *concurrency relation* $\| \subseteq X \times X$ contains the pairs (x_1, x_2) such that $M \geq M_{x_1} + M_{x_2}$ for some reachable marking M. In particular, two transitions t_1, t_2 belong to the concurrency relation if they can occur concurrently from some reachable marking, and two places belong to the concurrency relation if they are simultaneously marked at some reachable marking.

 The concurrency relation is very related to the *co* relation used in the theory of nonsequential processes [BF]: (x_1, x_2) belongs to the concurrency relation if and only if some nonsequential process of (N, M_0) contains two elements in *co* labeled by x_1 and x_2. This is in fact a more elegant definition, but, since it requires the introduction of a number of concepts, we use the one above.

 We now define the structural concurrency relation, first presented in [K1]. Let (N, M_0) be a system, where $N = (S, T, F)$, and let $X = S \cup T$. The *structural concurrency relation* $\|^A \subseteq X \times X$ is the smallest symmetric relation such that:

(i) $\forall s, s' \in S$: $M_0 \geq M_s + M_{s'} \Rightarrow (s, s') \in \|^A$

(ii) $\forall t \in T$: $(t^\bullet \times t^\bullet) \setminus id_T \subseteq \|^A$

(iii) $\forall x \in X \, \forall t \in T$: $\{x\} \times {}^\bullet t \subseteq \|^A \Rightarrow (x, t) \in \|^A \wedge \{x\} \times t^\bullet \subseteq \|^A$

where id_T denotes the identity relation on T.

Theorem 5.1 [K1]
  (a) For every net system, $\| \subseteq \|^A$.
  (b) For live T-systems, $\| = \|^A$. □

6.   THE MAIN RESULT

We prove in this section that $\|^A$ and $\|$ coincide for Regular systems (and therefore for Regular STGs). The reader which is not interested in the details of the proof can safely jump to the next section. The proof has much in common with the proof of the Second Confluence Theorem [DE], which we now recall.

6.1   THE SECOND CONFLUENCE THEOREM

The Second Confluence Theorem [DE] states that if two live markings M_1 and M_2 of a Regular system agree on all S-invariants, then they have a common successor, i.e., there exists a marking that is reachable from both M_1 and M_2. Since it can be easily shown that any two reachable markings agree on all S-invariants, it follows that any two reachable markings have a common successor. The result can be generalized to a set $M_1, \ldots, M_n$ of markings which agree pairwise on all invariants, and in the sequel we consider this more general version.

Let us first recall the notions of markings that agree on all S-invariants. Two markings M and M' of N *agree* on all S-invariants if $I \cdot M = I \cdot M'$ for every S-invariant I of N.

Theorem 6.1 [DE] Let (N, M_0) be a system, and let M be a reachable marking. Then M and M_0 agree on all S-invariants. □

The proof of the Second Confluence Theorem distinguishes two cases, according to whether the EFC-net N is a T-net or not. The first case is easily solved using the following result, which states that for T-systems the converse of Theorem 6.1 holds:

Theorem 6.2 [DE] Let (N, M_0) be a live T-system. A marking M is reachable iff it agrees with M_0 on all S-invariants. □

Since $M_1, \ldots, M_n$ are live and bounded and agree on all S-invariants, they are all reachable from each other. Therefore, any of them is a common successor of all the others.

In the second case, when N is not a T-net, the proof makes use of a reduction procedure given in the previous sections. N is split into two: a CP-subnet $\widehat{N} = (\widehat{S}, \widehat{T})$, and the subnet $\overline{N}$ generated by all the nodes that do not belong to $\widehat{N}$.

Theorem 3.4 guarantees that N can be split.

Once N is split, we let n particular sequences occur from $M_1, \ldots, M_n$. These sequences contain only transitions of $\widehat{T} \setminus \widehat{T_i}$.

Proposition 6.3 [DE] Let N be an EFC-net, and let $M_1, \ldots, M_n$ be live and bounded markings of N that agree on all S-invariants. Let $\widehat{N}$ be a CP-subnet of N, let $\widehat{T_i}$ be the set of way-in transitions of $\widehat{N}$, and let $\overline{N} = N \setminus \widehat{N}$. There exist occurrence sequences $M_1[\sigma_1\rangle M_1', \ldots, M_n[\sigma_n\rangle M_n'$, where $\sigma_1, \ldots, \sigma_n$ contain only transitions of $\widehat{T} \setminus \widehat{T_{in}}$, such that

(a) No transition of $\widehat{T} \setminus \widehat{T_{in}}$ is enabled at $M_1', \ldots, M_n'$,

(b) $\widehat{M_1'} = \cdots = \widehat{M_n'}$, where $\widehat{M}$ denotes the projection of M onto the places of $\widehat{N}$,

(c) $\overline{M_i} \leq \overline{M_i'}$ for $1 \leq i \leq n$, where $\overline{M}$ denotes the projection of M onto the places of $\overline{N}$, and

(d) $\overline{M_1'}, \ldots, \overline{M_n'}$ are live and bounded markings which agree on all S-invariants of $\overline{N}$.

□

Using Lemma 3.2, Proposition 6.3 can be easily generalized to Regular systems. After the occurrence of these sequences we 'freeze' the transitions of the CP-subnet, i.e., we forbid them to occur again, and so preserve the equality $\widehat{M_1'} = \cdots = \widehat{M_n'}$. If $\overline{N}$ is a T-net, then Theorem 6.2 can be applied, and we are done. Otherwise, by Theorem 3.4 and Proposition 6.3c we can iterate the procedure until we get two markings which coincide everywhere, and are therefore the same. This marking is a common successor of $M_1, \ldots, M_n$. Instead of freezing the transitions of the CP-subnet, we can equivalently remove them and consider thereafter the remaining net $\overline{N}$.

6.2   THE NEW RESULT

In order to adapt these results to the concurrency problem for Regular systems, we take a closer look at the proof of Theorem 6.1. The proof is based on the notion of T-component.

It is easy to see that if a net is T-coverable then every node of the net, and not only every transition, belongs to some element of a cover.

Now, in order to find a CP-subnet, we proceed as follows. We take a minimal cover $\mathcal{C}$ of N by T-components. Since N is not a T-net, we have $|\mathcal{C}| > 1$. We construct the (non-directed) graph $G = (V, E)$ as follows. V is the set $\mathcal{C}$ and E is the set of pairs (N_i, N_j) such that N_i and N_j have at least one common node. The graph G is connected because $\mathcal{C}$ is a cover of N and N is connected. Moreover, G has at least two nodes because $|\mathcal{C}| > 1$.

We choose a spanning tree of G, and select one of its leaves, say N_1. We then construct a maximal set of nodes X of N_1 satisfying the following properties: (a) the net generated by X is connected, and (b) no element of X belongs to a T-component of $\mathcal{C} \setminus \{N_1\}$. The set X is nonempty, because $\mathcal{C}$ is a minimal cover. The subnet N_X generated by X is a CP-subnet.

We prove a preliminary result, and then our main theorem.

Proposition 6.4 Let (N, M_0) be a Regular system, and let s and t be a place and a transition of N such that ${}^\bullet t = \{r_1, \ldots, r_n\}$. Assume that for every $1 \le i \le n$ there exists a reachable marking M_i such that $M_i \ge M_s + M_{r_i}$. Then there exists a reachable marking $M \ge M_s + \sum_{i=1}^{n} M_{r_i}$.

Proof: in [K3]. $\square$

Proposition 6.4 leads to our main result:

Theorem 6.5 The relations $\|$ and $\|^A$ coincide for Regular systems.

Proof: We have $\| \subseteq \|^A$ by Theorem 5.1a. We prove that the $\|$ relation of a Regular system (N, M_0) satisfies the three conditions of Definition of the structural concurrency relation. Since $\|^A$ is the smallest symmetric relation satisfying these conditions, we have $\|^A \subseteq \|$, which finishes the proof. Condition (i) follows easily from the definition of $\|$. Condition (ii) is a direct consequence of the liveness of (N, M_0). Condition (iii) follows immediately from Proposition 6.4. $\square$

7. COMPUTING THE STRUCTURAL CONCURRENCY RELATION

In [K1], we present a $O(n^5)$ algorithm for the computation of $\|^A$ in an arbitrary net system, where n is the number of places and transitions of the net. In [KE] we shown that $\|^A$ can be computed in $O(n^4)$ time, and in $O(n^3)$ time for free-choice systems.

For self-containment of the article we repeat the algorithm given in [KE] for ordinary nets.

Algorithm 2

Input: A system (N, M_0), where $N = (S, T, F)$.
Output: $R \subseteq X \times X$.

begin
$$R := \{(s, s') \mid M_0 \geq M_s + M_{s'}\} \cup \bigcup_{t \in T} t^{\bullet} \times t^{\bullet};$$
$E := R \cap (X \times S);$
while $E \neq \emptyset$ **do**
        choose $(x, s) \in E;\ E := E \setminus \{(x, s)\};$
        for every $t \in s^{\bullet}$ **do**
            if $\{x\} \times {}^{\bullet}t \subseteq R$ **then**
                $E := E \cup ((\{x\} \times t^{\bullet}) \setminus R);$
                $R := R \cup \{(x, t)\} \cup (\{x\} \times t^{\bullet});$
            endif
        endfor
endwhile
end

Proposition 7.2 Algorithm 2 terminates, and after termination $R = \|^A$.

Proof: Observe that $E \subseteq R$ is an invariant of the while loop and holds initially. Therefore, each execution of the while loop removes from E an element of $E \cap R$. This element is never added to E again. So the algorithm terminates.

Let Q be the value of R after termination. We prove: (1) $Q \subseteq \|^A$.

We have $\{(s, s') \mid M_0 \geq M_s + M_{s'}\} \subseteq \|^A$ and $\bigcup_{t \in T} t^{\bullet} \times t^{\bullet} \subseteq \|^A$ by definition.

So initially $R \subseteq \|^A$.

Moreover, it follows easily from the definition of $\|^A$ that $R \subseteq \|^A$ is an invariant of the while loop. So we have $Q \subseteq \|^A$.

(2) Q satisfies the three conditions of the Definition of the structural concurrency relation.

Conditions (i) and (ii) follow immediately from the initialization of R. For condition (iii), let $x \in X$ and $t \in T$. We have to prove:

$$\{x\} \times {}^{\bullet}t \subseteq Q \implies (x, t) \in \|^A \wedge \{x\} \times t^{\bullet} \subseteq Q$$

If $\{x\} \times {}^{\bullet}t$ is not included in Q, we are done. So assume $\{x\} \times {}^{\bullet}t \subseteq Q$.

Let (x, s) be the last element of $\{x\} \times {}^{\bullet}t$ which is removed from E during the execution of the algorithm. As we have seen above, (x, s) is never added to E again.

Assume that immediately after (x, s) is removed from E, we have $(x, s') \notin R$ for some $s' \in {}^{\bullet}t$. We prove that (x, s') is never added to R

later on. Every new element added to R is also added to E, and every element of E is removed before termination. Therefore, if (x, s') were added to R it would later be removed from E, which contradicts our assumption about (x, s).

Since $\{x\} \times {}^{\bullet}t \subseteq R$ and no element of $\{x\} \times {}^{\bullet}t$ is added to R after (x, s) is removed from E, we already have $\{x\} \times {}^{\bullet}t \subseteq R$ immediately after (x, s) is removed from E. Then, the next execution of the for loop adds $(\{x\} \times t^{\bullet})$ to Q. So $(\{x\} \times t^{\bullet}) \subseteq Q$ after termination.

$Q = \|^A$ follows from (1), (2) and the minimality of $\|^A$. $\square$

We calculate the complexity of the algorithm when the subsets $X \times X$ (in particular, the incidence relations of the net) are encoded as bidimensional arrays $X \times X \rightarrow \{0, 1\}$. In this case, the algorithm needs $O(|X|^2)$ space.

The initialization of Q, E and N takes $O(|S|^2 \cdot |T|)$ time. The while loop is executed at most $O(|S| \cdot |X|)$ times, because each iteration removes one element from E which is never added to it again. In each iteration the execution of the for loop takes $O(|S| \cdot |T|)$ time ($O(|T|)$ iterations with $O(|S|)$ time each). So the algorithm runs in $O(|S|^2 \cdot |T| \cdot |X|)$ time.

8.   CONCLUSIONS

We have presented an $O(n^4)$ algorithm for the computation of the concurrency relation of Regular systems, where n is the number of nodes of the net. Our work was motivated by the interesting applications of the concurrency relation to the design and verification of asynchronous circuits. Our algorithm can be used to detect inconsistently encoded Regular STGs. It can also be used as subroutine in the algorithm for checking the Unique State Coding property used in [PC].

Our paper adds one more to the list of results on the concurrency problem, i.e., the problem of deciding if two given transitions are concurrently enabled at some reachable marking.

Our algorithm also can be used to solve the 1-boundedness problem: decide if a given Regular system is 1-bounded. It follows easily from the definition of the concurrency relation that a net system is 1-safe iff its concurrency relation is irreflexive. So the 1-boundedness problem can be solved in $O(n^4)$ as well. This improves the complexity of earlier algorithms based on linear programming.

We improve the reduction method given in [E1]. First, we generalize this method to Regular Petri nets. Initially this method was used only for Extended Free Choice Petri nets. Second, we give more precise evaluation of the complexity of this method ($O(|S| \times |T|)$), versus polynomial without specifying the degree of the polynomial. Third, we improve the

Reduction Rule. We eliminate one computationally expensive step and reduce overall complexity of the reduction algorithm.

References

[AFN] P.Alimonti, E.Feuerstein, U.Nanni. "Linear time algorithms for Liveness and Boundedness in Conflict-free Petri nets", *Proc. of 1-st Latin American Symposium on Theoretical Informatics*, LNCS 583, (1992).

[B] E.Best. "Structure theory of Petri nets: the free choice hiatus". *Advances in Petri Nets 1986*, LNCS 254. (1987) 168-205.

[BF] E.Best, C.Fernández. "Non-sequential processes. A Petri net view". EATCS Monographs on Theoretical Computer Science, 13, 1988.

[CEP] A. Cheng, J. Esparza, and J. Palsberg. "Complexity Results for 1-safe Nets". *Theoretical Computer Science*, 147:117,136, 1995.

[C] T.A. Chu. "Synthesis of Self-timed VLSI Circuits from Graph-theoretic Specifications". Ph.D. thesis, MIT, 1987.

[D1] J.Desel. "Regular marked Petri nets". *Proc. of the 19th Int. Workshop on Graph-Theoretic Concepts in Computer Science/ Jan van Leeuween (ed.)* LNCS 790, pp.264-275 (1993).

[D2] J.Desel. "A proof of the Rank Theorem for extended free choice nets". *Proc. of the 13th Int. Conf. on Appl. and Theory of Petri nets*. Sheffield, June 1992. LNCS 616.

[DE] J.Desel, J.Esparza. "Free Choice Petri Nets", Volume 40 of Cambridge Tracts in Theoretical Computer Science. Cambridge University Press, 1995. 244p.

[E1] J. Esparza. "Reduction and Synthesis of Live and Bounded Free Choice Petri Nets". *Information and Computation*. Vol.114. No.1. 1994.

[E2] J. Esparza. "A Solution to the Covering Problem for 1-Bounded Conflict-Free Petri Nets Using Linear Programming." *Information Processing Letters*, 41:313,319, 1992.

[ES] J.Esparza, M.Silva "On the Analysis and Synthesis of Free Choice Systems". *Advances in Petri Nets*. 1990. G.Rosenberg (Ed.). Lecture Notes in Computer Science 483.

[FO] O.Fergemand, A.Olsen. "Introduction to SDL-92". *Computer Networks and ISDN Systems*, (26): 1143-1167, 1994.

[Ha] M.T.Hack. "Analysis of production schemata by Petri nets", TR-94, MIT, Cambridge, MA, 1972, pp.119. Corrections 1974.

[Ho] C.A.R.Hoare. "Communicating Sequential Processing", Prentice/Hall, London, (1985).

[KPCR] A. Kondratyev E. Pastor, J. Cortadella, and O. Roig. "Structural algorithms for the synthesis of speed-independent circuits from signal transition graphs". *Proceedings of the European Design and Test Conference*, 1996.

[K1] A.Kovalyov. "Concurrency Relations and the Safety Problem for Petri Nets", *Proc. 13th Int. Conf. on Application and Theory of Petri nets, Sheffield, UK*, LNCS 616, pp. 299-309, June 1992.

[K2] A.Kovalyov, "An $O(|S| \times |T|)$-Algorithm to verify if a net is regular", *Proc. 17th Int. Conf. on Application and Theory of Petri nets, Osaka, Japan*, LNCS 1091, pp. 366-379, June 1996.

[K3] A.Kovalyov, "A polynomial algorithm to compute the concurrency relation of regular Signal Transition Graphs", *Second Workshop "Hardware Design and Petri nets" within 20th International Conference on Application and Theory of Petri Nets (ICATPN'99)*, Williamsburg, Virginia, USA, June 1999.

[KE] A.Kovalyov and J.Esparza, "A polynomial algorithm to compute the concurrency relation of Signal Transition Graphs", *Proc. 3rd Workshop on Discrete Event Systems (WODES'96), Edinburgh, Scotland, UK*, August 1996.

[KL] A.Kovalyov and R.McLeod. "New Rank Theorems for Petri Nets and their Application to Workflow Management", *1998 IEEE International Conference on Systems, Man, and Cybernetics*, San Diego, California, USA.

[Kem] P.Kemper. "$O(|P| \times |T|)$-algorithm to compute a cover of S-components in EFC-nets". Forschungsbericht 543, Universität Dortmund, 1994.

[LTS] P.E. Lauer, P.R.Torrigiani and M.W. Shields. "COSY - a System Specification Language Based on Paths and Processes." *Acta Informatica*, 12, 1979, pp.109-158.

[LW] K. Lautenbach and H. Wedde. "Generating Mechanisms by Restrictions". LNCS 45, pp.416-422, 1976.

[LS] L. Lavagno and A. Sangiovanni-Vincentelli. "Algorithms for synthesis and testing of asynchronous circuits". Kluwer Academic Publishers, 1993.

[M] T.H.Y. Meng, editor. "Synchronization Design for Digital Systems". Kluwer Academic Publishers, 1991.

[Mi] R. Milner. "A Calculus of Communicating Systems". LNCS 92, 1980.

[MR] G.Memmi, G.Roucairol. "Linear Algebra in net Theory". *Net Theory and Applications*, LNCS 84, 213-223. 1980.

[Mu] T. Murata. "Petri nets: properties, analysis and applications". *Proceedings of the IEEE*, vol. 77, n. 4, 1989, p.541-580.

[PC] E.Pastor and J.Cortadella. "An efficient unique state coding algorithm for signal transition graphs". *Proceedings of the IEEE International Conference on Computer Design: VLSI in Computers and Processors*, pages 174-177, 1993.

[Ta] R. Tarjan. "Depth-first search and linear graph algorithms". *SIAM J. Comp.*, Vol.1, p.p.146-160.

[Ya] Yakovlev, A.V. "On limitations and extensions of signal transition graph model for designing asynchronous control circuits". *Proc. Int. Conf. on Computer Design (ICCD'92)*, Cambridge, MA, October 1992, IEEE Comp. Society Press, N.Y., pp. 396-400.

[Ye] H. C. Yen. "A Polynomial Time Algorithm to Decide Pairwise Concurrency of Transitions for 1-Bounded Conflict Free Petri Nets". *Information Processing Letters*, 38:71-76, 1991.

III

THEORY AND PRACTICE OF PETRI NET BASED SYNTHESIS

Chapter 7

SYNTHESIS OF SYNCHRONOUS DIGITAL SYSTEMS SPECIFIED BY PETRI NETS

Norian Marranghello
Department of Computer Science
São Paulo State University
Rua Cristóvão Colombo, 2265
15054-000 Rio Preto, SP, Brazil
norian@dcce.ibilce.unesp.br

Jaroslaw Mirkowski
Technical University of Zielona Gora
16 Trasa Polnocna
65-246 Zielona Gora, Poland
j.mirkowski@pz.zgora.pl

Krzysztof Bilinski
ADB Polska Ltd
50, Podgorna Street
65-119 Zielona Gora, Poland
k.bilinski@adb.pl

Abstract   During the last fifteen years the importance and use of Petri nets as a language for modeling digital systems have greatly increased. Many new computer-aided design tools have been developed that employ Petri nets in the analysis, verification, and synthesis of this sort of hardware. This chapter aims at presenting an overview of the research going on the application of Petri nets to the description of digital systems and the synthesis of hardware from these descriptions.

Keywords: Hardware/Software CoDesign, Petri Nets, Synchronous Digital Systems

A. Yakovlev et al.(eds.), Hardware Design and Petri Nets, 129-150.
© 2000 Kluwer Academic Publishers.

1.   INTRODUCTION

The automation of the synthesis process of digital systems is an issue whose importance steadily grows with the constant increase in the complexity of such systems. The use of Petri nets in the modeling and simulation of complex systems has proved very worthwhile. Many groups around the world have already successfull y used Petri nets for the specification, analysis, and synthesis of digital systems for several years. The main goal of this chapter is to provide an overview of the research going on the synthesis of digital systems from Petri net descriptions.

The first step in the design of a digital system is its specification. The very first ideas are usually presented in a natural language. The task of the designer is to develop the idea into a collection of hardware and/or software components to perform the required function(s).

Kishinevsky et al. [40] present an anecdote that, in our opinion, illustrates very well the vagueness of such kind of specification. Their illustrative story is about the following: A soldier in the front comes to his commanding officer with the idea of building a weapon to reach far beyond the enemies' lines so as to cut their supplies. Weakened, the opposing military forces would more easily be defeated. The soldier was immediately sent to the nation's president. When asked how to build such a magnificent weapon the soldier replied that his duty was to suggest the idea. Likewise, in order to develop it into a practical device there was a full team of designers much more qualified than him. Although Kishinevsky and co-workers offer a slightly different version of the story, using it in a somewhat different context too, it serves very well to show the level of abstraction in which an initial specification can be proposed. The designer can be presented with specifications such as the one above and has to eventually come up with a system to carry out the required set of functions.

Considering such broadness it can be argued that the steps leading from an initial specification towards a final implementation of a system may be thought of as a process of synthesis. This is certainly the case that the scope of the terms "final implementation of a system" and "process synthesis" is too coarse. Thus, more specific definitions are needed. These two terms can in fact be spelled out with the aid of the following three words: system, implementation and synthesis.

Throughout this chapter the word system will refer to a digital system, i.e., such a system that manipulates discrete elements of information, unless strictly stated otherwise.

The second word, implementation, can have two different scopes, i.e., it is used both in the software and hardware contexts. In the former, it is assumed to be the generation of a piece of code written in some computer programming language. This code should be ready to use, if a low-level language is employed, or directly convertible to a machine-understandable language, when the code is produced in a high-level programming language. In case of hardware it is interpreted in this chapter as the actual generation of an integrated circuit directly into silicon or some other suitable substrate.

The third word, synthesis, needs a bit more elaborate and longer explanation, for which some help from Gajski and co-workers is greatly welcome. In this chapter it is adopted the more general definition of synthesis as the process of refining a description given in some higher level of abstraction into another one at a lower level of abstraction. Considering the above definition of synthesis and the division of the design space as proposed by Gajski and Kuhn [29], it is possible to define high- and low-level synthesis as being "a translation from a behavioral description into a structural description" [30], and a translation from a structural description into a geometrical one, respectively.

It is possible to identify several levels of abstraction within each region of the design space in which to describe a digital system. From a more theoretical point of view down to a more concrete one at least four levels of abstraction can be distinguished within structural region, namely the architectural, register-transfer, logic, and circuit levels. More details on each such level are given in the following paragraphs.

The state-of-the-art approach to synthesize digital systems consists in the utilization of the so-called co-design techniques. In this case the design process usually starts with a behavioral specification of the system. The initial specification is progressively refined in order to obtain the best match between its software and hardware components. After a series of refinements, the synthesis of the system is concluded with the implementation of both its software and hardware constituent parts.

This chapter is focused on the synthesis of hardware. The reader interested in synthesis of software or in more details about co-design is referred to the relevant literature [18, 24, 33, 49, 51, 75, 79].

As for the hardware part, the behavior of the system is initially provided in a suitable hardware description language. At the beginning of the design process the system is described from such a behavioral perspective. Then, a transformation from the behavioral to a structural viewpoint, frequently at the architectural level, where the system is described as a set of processors, memories, controllers and interfaces, is carried out. Thereupon, this description is gradually refined to lower levels of abstraction. As a rule, the architectural description of the system is further detailed down to the register-transfer level, when the system is split into a data path and a control unit. The data path is composed of a set of registers to store information and some functional units to manipulate the data while transferring them among the registers. The control unit is a combinational circuit used for sequencing the actions in the data path. At this point, the description of the system consists of an interconnection of registers, multiplexers and some other high-order logic entities, including some time dependence among them as well. At this stage, the so-called high-level synthesis process is completed. This is also, the level of detail in which most existing co-design systems provide the description of the hardware.

Next, an intermediate step converting the register-transfer description into logic level one is needed. At the logic level the structure of the system is described in terms of logic gates and their timing relationships. It is considered here as a changeover procedure because it is not clear at the moment whether this is part of a high- or low-level synthesis system. Existing synthesis approaches, at either level, may include or not this kind of transformation. Anyhow, the result is a description of the system in terms of (possibly simple) logic gates and a reasonably detailed timing relationship among such entities.

Typically, during the low level synthesis process the gates constituting the logic level description of the system are converted to transistors, whose parameters, e.g., sizes and types, are computed taking into account the timing information available. From the circuit level structural description a geometric one is obtained. Similar to the structural region of the design space, the geometric one can be seen at several levels of abstraction. Nevertheless, let's assume here that the resulting geometrical description be a layout of the final circuit, containing a profile of the geometric shapes to be used in the manufacturing of the desired system. Each of these shapes, such as rectangles, squares, etc., represents a particular type of substrate and was conveniently sized in order to reflect the characteristics of each transistor in the proposed hardware.

The specification of synthesized hardware is usually given in a specialized description language (Hardware Description Language, HDL). There are several HDLs available, but the majority of solutions nowadays are based either on VHDL or Verilog HDL specifications. Nevertheless, Petri nets are becoming continually more popular for the description of digital systems. This is especially true in the case of asynchronous circuits, where a great likeness of some fundamental principles to those of Petri nets can be observed. During the last 15 years or so the importance and use of Petri nets as a design aid to digital systems have greatly increased. Three main reasons can be identified as catalysts of such a trend. First, Petri nets constitute a language capable of capturing causality relations, concurrency of actions and conflicting conditions from digital systems in a natural and convenient way. Second, Petri nets embed a theory that allows the description, analysis and verification of digital systems in a formal yet easy to use basis. And last but not least, the evolution of computer hardware and software brought to reasonably cheap personal computer such a computing power that until few years ago was available on expensive workstations only. This in turn makes computationally intensive operations on Petri nets feasible.

The rest of this chapter is organized as follows. In section two a brief review of the main Petri net related techniques used during the design of digital systems is provided. In section three an overview of the work going on the subject title is presented. Our overview concentrates mostly on synchronous domain as asynchronous hardware design based on Petri nets has recently been thoroughly investigated elsewhere [84]. Some conclusions are offered in section four. Finally, the "References" section contains an extensive list of publications in the field.

2. DESIGN TECHNIQUES

In this section it is offered an overview of the main Petri net related techniques used during the design of digital systems. The reader may find some resemblance with the expositions by Yakovlev et al. [84, 85]. This is indeed the case as those papers positively influence this section. It is included here for two reasons: first to serve as a reference for the reader not quite familiar with some specific terms, second for the sake of completeness of the present overview.

It is assumed that the reader has a basic knowledge both of Petri nets and of hardware. Those unfamiliar with Petri nets could have profit from the paper by Murata [55] and the books by Reisig [74] and by Jensen and Rozenberg [37]. Those unfamiliar with hardware are referred to the books by Gajski et al. [30] and by Kishinevsky et al. [40].

The design process of a digital system comprises roughly three parts: modeling, analysis/verification, and synthesis; each of which is informally explained below.

The formal verification of a system requires it to be described through a formal model. The modeling language considered here is Petri net, which is a bipartite directed graph, i.e., a graph with two kinds of vertices: local states (usually referred to as places) of the net, graphically represented by circles or ellipses, and actions (usually referred to as transitions) of the net, graphically represented by rectangles. Each transition is connected to one or more input places through directed edges, which have their origin at a place, and end at the considered transition. Each transition is also connected to one or more output places through directed edges, which originate at the transition, and end at the output place(s). The global state of the system (also known as net marking) is represented by the assignment of tokens to each of the net places, resulting in a general net marking. A change to the net marking can be seen as the variation of the token assignments, i.e., by moving the tokens around through the places. A change in the net marking corresponds to the execution of actions (known as transition firing or occurrence) according to the following basic rules:

(a) a transition is enabled if each of its input places has at least one token;

(b) any enabled transition can occur, and its firing is represented by removing a token from each of the corresponding input places and inserting a new token in each of its output places; and

(c) enabled transitions can occur concurrently as long as they are independent, i.e., use different tokens.

According to such a definition, a Petri net can be represented by a tuple $PN = (P, T, F, Mo)$ where P is the set of places, T is the set of transitions, F is a flow relation defining directed edges connecting either places to transitions or transitions to places, and Mo is the initial marking of the net. Next, a labeled Petri net is defined as a Petri net in which every transition is labeled with a symbol from an alphabet A, originating the corresponding labeling function $L : T \mapsto A$. In order to bring this theory closer to digital systems a signal transition graph (STG) is defined as a Petri net whose transitions are labeled with the names of binary signal transitions, i.e., the alphabet considered is the set $A = \{a+, a-, a \sim\}$. The elements of the alphabet correspond to the upward $(a+)$, downward $(a-)$ and either $(a \sim)$ transition of signal a. Then, an STG can be viewed as a triple $STG = (PN, A, L)$, where its components are defined as above.

The model produced must be analyzed, i.e., the designer needs to answer some questions about the behavior of the system. Furthermore, it is desirable to verify some characteristics of the system; or putting it in another way, it is desirable to rigorously prove that the designed system possesses some formally stated attributes. In order to do so, three approaches are considered, namely: reachability graphs, symbolic traversal, and partial orders.

Reachability analysis uses a graph containing all possible markings of the net and all possible sequences of firings. The main problem with this method is that the number of possible states grows enormously even for small examples, causing the state explosion. To avoid this problem Valmari [82] proposed a "stubborn sets" method. In this method only one of a group of concurrently enabled transitions is analyzed at a time. In order to overcome the state explosion problem, only part of the reachability graph is generated. Consequently, not all properties of the system can be analyzed. For instance, Valmari proved that his method could always detect deadlocks in the system. However, by hiding some states of the system, it may not be possible to prove the Complete State Coding (CSC) property [19], thus implying that it may not be possible to verify whether the logic circuit corresponding to the modeled system can be implemented. Furthermore, as the states of the system are manipulated and not all of them are generated, this procedure may conceal the causal ordering of events preventing a timing analysis of the model.

Another approach starts by constructing a k-variable Boolean function where each variable represents one place of the net and expresses the binary condition of existence or not of a token in the corresponding place. In such a way the net is described through a characteristic function, which in turn is represented as a binary decision diagram (BDD) [17]. A transition function, transforming markings into one another, is computed for each enabled transition at each marking. By successively applying the transition function to each marking, the symbolic image of the full state-space is generated. As compared to the reachability analysis, the symbolic traversal approach has a better performance due to Boolean characterization. Covering all reachable markings it is also possible to prove the CSC property with this technique. Nonetheless, there still may be not enough information about the causality ordering of events in order to proceed a timing analysis of the model.

The partial ordering representation is given by the Petri net unfolding [50, 26]. Informally, the unfolding of a net is its transformation into an equivalent one in which the cycles are broken and extra places are inserted to preserve the semantics of the original net. With this approach, there is enough information to cover all possible markings. As the cycles in the original net are broken, it could be the case that the information about the causal relationship of events became hidden by such cuts. Though, Semenov and Yakovlev [77] used it for the analysis of timed models of asynchronous circuits. In order to do that, time-state classes are associated to the markings produced allowing the timing analysis.

The third part of the design process is the actual synthesis of the desired system. Two ways are possible here, namely the syntax directed and the STG-based approaches.

The syntax directed synthesis starts with a description of the system as a labeled Petri net. The components of this net are effectively replaced with circuit elements. For this to be feasible the net must be at least bounded, if not 1-safe, so that the generated circuit have a meaningful implementation. There are two ways to generate the required circuit, they are the place-to-latch and the event-based mappings. The former maps each place in the net to a memory latch (flip-flops in the case of 1-safe nets and up/down counters for bounded ones) and each transition to some appropriate random logic at the input of the corresponding latches. This can be used to synthesize relatively large circuits as long as the constraints on area and speed are not very critical. The latter is an extension of Patil's mapping. He used six elements and eight modules composed of these elements to represent Petri net structures [66, 67]. This extension, uses a seventh component, named decision-wait that allows the synchronization of events in different groups of mutually exclusive signals. This gives way to the mapping of unsafe, although bounded, Petri nets. Syntax directed approaches require that all transformations be done at the Petri net level. The reason for this is that doing any change at the circuit level would at least risk destroying the semantic soundness of the circuit in respect to the verified net.

A complimentary technique takes the system specification in terms of an STG, and generates its state graph. The state graph of an STG is the binary encoded form of the reachability graph of the corresponding labeled Petri net. Using Boolean minimization techniques a Boolean function is derived from the state graph. Thus, the desired circuit can be generated through one of a number of methods available [42, 76].

3.   APPLICATIONS

The key part of this overview is a description of approaches that relate to using Petri nets in hardware synthesis. They can be divided into three main groups, discussed in the following subsections: (1) asynchronous hardware, (2) synchronous parallel controllers, (3) high-level synthesis and hardware-software co-design.

3.1   ASYNCHRONOUS HARDWARE

Synthesis of asynchronous digital systems from Petri net descriptions has been a topic of intensive research in several universities worldwide since the end of the eighties [84]. Main works in the field are conducted now in the universities of Aizu (Japan), Barcelona (Spain), Newcastle (UK) and Torino (Italy). An overview of asynchronous systems synthesis from Petri nets is extensively described elsewhere [84], we will concentrate ourselves on synchronous systems only.

3.2 SYNCHRONOUS PARALLEL CONTROLLERS

3.2.1 Parallel controller synthesis - methods.
The problem of synthesizing synchronous parallel controllers from Petri net specifications has been under investigation for a long time. This work has led to a number of methods that can be classified into four categories: direct implementation, place encoding, marking encoding, and net decomposition. The first three are fairly independent and employ different techniques. The last method is often an introductory stage, which is followed by the use of one of the encoding approaches.

3.2.1.1 Direct implementation.
A method of directly synthesizing parallel controllers from a Petri net was presented by Patel [65]. The method is based on the proposal of one-to-one implementation of a Petri net description, which is a straightforward extension of the one-hot-code approach used for realizing FSMs [35]. In the method each place is implemented with a D-type flip-flop. Transitions are realized using two level combinational logic with an AND gate for each transition and an OR gate for collecting transitions which transfer a token to a particular place (Figure 7.1). The initial marking of a Petri net is realized by presetting flip-flops that represent places holding tokens in a Petri net specification. In some technologies, where there is a global reset on a powering-up (e.g. Xilinx LCAs), an additional inverter is associated with the output of every flip-flop representing an initially marked place.

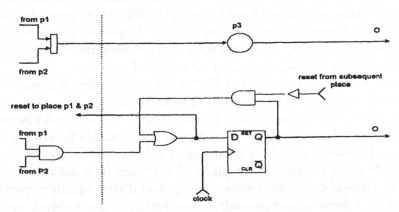

Figure 7.1 Implementation of a place and its input transition by the direct method.

The method results in a simple combinational logic and a state register of a size equal to the total number of places. This approach, although very simple in use, is wasteful of flip-flops, and so for large controllers may be inefficient. However, in implementations in which the number of flip-flops is not really a limitation and where each flip-flop is associated with a limited combinational capability it may be useful (e.g. when using FPGAs). This approach usually leads to circuits with potentially high clocking rates. There are several examples of the application of the above concepts to controller design [16, 44, 59, 78].

3.2.1.2 Place encoding.   Place encoding is analogous to state assignment for FSMs. The number of flip-flops needed to implement a parallel controller depends on the way they are assigned to encode places in a Petri net. The silicon area occupied by the controller is usually a tradeoff between the number of flip-flops comprising the state register and the number of gates comprising the combinational logic. An optimal solution is obtained if places share the same multiple-flip-flop state register. Since different intermediate states can be active simultaneously, constraints on the use of the register are necessary. The state assignment constraints can be formulated as follows [1]:

- The intermediate states that are concurrent (i.e., places which hold tokens simultaneously) must have non-orthogonal codes.

- The intermediate states that are non-concurrent can have orthogonal codes.

These rules may be rephrased so as to explicitly describe the use of flip-flops [64]: intermediate states that are never active at the same time can share all of their state variables, but intermediate states that can be active simultaneously must have codes that differ by at least one state variable to distinguish them from each other. Moreover, if intermediate states that can be active simultaneously share some state variables, then their codes for these variables must be identical.

Adamski [1], and Amroun and Bolton [3] independently presented a method for encoding Petri net places. The method uses a transformation of an original Petri net into a more general description, called a macronet [3, 7, 55, 63]. All arcs connected to fork or join transitions are cut and a set of sequential sub-nets is produced. Each sub-net is then replaced by a single place called a macroplace; this results in a macronet. The encoding method could not however be applied if structurally sequential places held tokens simultaneously. Generating a concurrency matrix and then performing the encoding may circumvent the problem. The concurrency matrix $C_{n x n} : n \in \aleph^*$ of a Petri net is defined as follows[64]:

$C(i, j) = 1$ if p_i and p_j never hold tokens simultaneously or when $i = j$,

$C(i, j) = 0$ otherwise.

The concurrency matrix is constructed by examining the reachability graph of the Petri net. An algorithm that optimizes the use of flip-flops regardless of whether the net has been previously reduced or not has been proposed by Kozlowski et al. [44]. The place encoding approach results in a decreased number of flip-flops and an increase in the complexity of the combinational logic in comparison to the direct implementation method. Many examples of the implementation of the place encoding method are presented in the literature [2, 13, 14, 36, 44, 61, 64]. Further reduction of the needed number of flip-flops can be achieved by applying a marking encoding approach.

3.2.1.3 Marking encoding. State transition graphs are semantically equivalent FSMs obtained from the Petri net behavioral analysis, so that each state represents a marking of the Petri net [62]. The task of generating a state transition graph from a Petri net is very similar to the task of generating a Petri net's reachability graph. Every marking of the net describes one state in the state transition graph. State transitions are obtained by generating combinations of input signals that either enable some transitions in the Petri net to fire, so that the net marking changes, or disable all transitions in the net from firing, so that the marking stays unchanged. When generating a state transition graph the entire interpretation of a Petri net is taken into account. For each marking and every tested input, the marking encoding implements a Petri net as a sequential controller. Figure 7.2 shows a synchronous interpreted Petri net and a state transition graph, which describes a functionally equivalent FSM. The dotted arc represents an enabling arc and dotted arc with an empty circle represents an inhibitor arc.

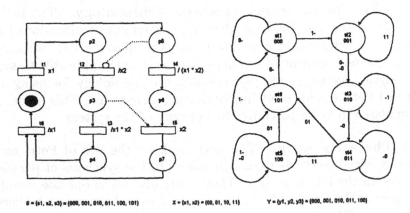

$S = \{s1, s2, s3\} = \{000, 001, 010, 011, 100, 101\}$ $X = \{x1, x2\} = \{00, 01, 10, 11\}$ $Y = \{y1, y2, y3\} = \{000, 001, 010, 011, 100\}$

Figure 7.2 Petri net and equivalent state transition graph.

This method results in the minimum number of flip-flops, but generates very complex combinational logic, so that the slowest and biggest designs are produced [7, 43, 45, 62, 69].

3.2.1.4 Net decomposition. The simplest method for Petri net decomposition into a set of sequential components is equivalent to the macronet construction method [6, 7, 45]. A component is generated by cutting all input and output arcs of the fork and join transitions. However, this tends to generate a large number of trivial sub-nets that contain only a few places. More advanced heuristic growth algorithms based on examining the structure of the net were presented as well [13, 14, 16, 56, 68, 79]. However, growth algorithms have a local scope; i.e., only a part of a Petri net that is adjacent to a component is considered. Another method uses a concurrency matrix to produce Petri net behavioral decomposition [60, 63]. The method is based on the prime implicant generation problem for FSMs [4]. Further, a decomposition approach based on the Petri net unfolding has been proposed [15].

An alternative approach for net decomposition into well-formed subnets, i.e., sub-nets with well-defined marking relation among places, was proposed [9]. In this approach control flow between a controller and a data path is analyzed and then partitions, such that each partition is closely associated with a set of data path elements, are identified. Recently, a BDD based approach to the state machine decomposition of Petri nets has been reported [31]. However, the method has exponential computational complexity with respect to the size of the net.

Usually each of the above approaches is followed by application of standard FSM based techniques to encode places in every component of the net.

3.2.2   Parallel controller synthesis - chronology. The problem of synthesizing parallel controllers from Petri net specifications has been under investigation for a long time. The first research work resulted in the development of methods that were only applicable to systems analysis. Azema et al. [8] presented a methodology for using an interpreted Petri net for modeling of concurrent systems. This model is then simulated in order to verify the behavior of the system.

3.2.2.1 The early work. The next work on the use of Petri nets resulted in initial methods that were suited to the synthesis of parallel controllers, in the late seventies. Those methods, which operate mainly on interpreted Petri nets, allowed synthesis of asynchronous controllers. Leung et al. [45] proposed two methods of realizing controllers using

Programmable Logic Arrays (PLAs). Interpreted Petri nets are used to represent a controller, input variables are only associated with transitions and output variables with places, so that only Moore-type controllers could be synthesized. The net was first decomposed into a set of sub-nets (FSMs) by cutting all join and fork transitions. Then, a flow table was associated with each of the sub-nets, or a reachability graph of each sub-net was generated and a unique binary code was associated with each node of the graph. Finally, both representations of a controller were implemented in PLAs.

Auguin et al. [6] presented a similar methodology for realizing controllers using PLAs. After the net decomposition, the sub-nets are implemented separately as a set of co-operating sub-devices. They introduced and extension to the previous methods, whose main idea is to generate a reachability graph of the entire net [7]. When the net became too complex to generate the reachability graph, the net was first simplified by reduction. It was suggested that a Petri net might be reduced by replacing all places between fork and join transitions by a single macroplace. In addition, they associated output variables with places and transitions, so that both Mealy and Moore controllers could be synthesized.

Toulotte and Parsy [81] proposed a method of realizing control structures with dedicated circuits. They used a heuristic algorithm to decompose the Petri net into a set of state machines, and then each of those sub-nets was directly implemented using one storage element per place.

New more efficient algorithms have been introduced since the mid-eighties. These algorithms are suited to the implementation of parallel controllers into VLSI technologies.

3.2.2.2 Modern synthesis methodologies.

Bruck et al. [16] presented a method of synthesis of complex controllers. A modified Petri net is used for the specification of the concurrent system - there are six types of transition, so six types of firing rules are used. Each of these transition types is implemented into its own hardware structure. The method allows the synthesis of asynchronous or partially asynchronous controllers, i.e. synchronous modules with asynchronous communication between each other. The entire specification is first partitioned into FSM modules, and then each of these modules is mapped to the CMOS implementation.

Peng and Kuchcinski [69] proposed an approach to implementing control structures by transforming their Petri net specification into microprograms, which is partially similar to the method described by Augin et al. [7]. They used extended timed Petri nets to specify synchronous parallel controllers. The main idea is that a reachability graph is first generated from the Petri net description, and then from this graph the microprogrammed controller is synthesized. This microprogram can then be coded directly into a control memory. The proposed method is supported by the CAMAD system [68]. CAMAD is an example of a Petri-net-based tool for design of digital circuits.

Balakrishnan et al. [9] presented a technique for synthesizing decentralized controllers based on analysis of interconnections between control unit and data path. They introduced a novel idea for partitioning the control structure. A modified Petri net is used to describe the controller's behavior, two additional signals are added to each transition and new rules for transition firing are presented. The proposed approach analyzed control flow between a controller and a data path and then identified partitions such that each partition is closely associated with a set of data path elements. Finally, each partition is implemented using PLA like structures. As a result the controllers have reduced area, increased speed and are more easily tested.

Nishimura and Zaky [56] proposed a synthesis methodology, which involves synchronous and asynchronous components. To model a concurrent system an extended Petri net is used, which contains three types of transitions. The Petri net specification is first decomposed into a set of state machines, with the restriction that all transitions in a sub-net must be asynchronous or synchronous with the same clock. Each of the sub-nets is then synthesized using standard FSM techniques. However, the set of state machines cannot cover the whole net – some individual nodes in the net can be synthesized directly as asynchronous control units.

Patel [65] introduced a method of direct implementation of a controller that is described using an interpreted synchronous Petri net. The controller's specification is first mapped to a generic hardware cell by applying a one-to-one approach. In the approach each transition is mapped onto an AND gate; each place is implemented with a D-type flip-flop and OR gate collecting its input transitions. After the net mapping the initial realization is minimized. Finally, technology mapping and optimization is performed. In the proposed approach the controllers were finally mapped into a Programmable Logic Array (PLA) or Logic Cell Array (LCA). In addition, a fast method of cost and speed estimation of the design is also presented.

Pardey and Bolton [62] presented a novel concept of synthesis of parallel controllers from Petri net description. The main idea is that a Petri net specification of a controller is transformed into a VHDL description of the controller, and then standard synthesis methods, which support

VHDL specification, are used. The emergence of VHDL as an industry standard favors its use.

3.3 HIGH-LEVEL SYNTHESIS AND HARDWARE-SOFTWARE CO-DESIGN

Intuitively, synthesizing controllers from Petri nets is quite a natural application, as the net itself represents possible states of the designed circuits and to the net nodes can be assigned both input and output signals. Assignment of a data path, however, poses a different challenge, as ordinary Petri nets do not support variables, types and operations that would be needed here. For that reason high-level synthesis based on Petri nets is not that well developed as the synthesis of controllers, but some results have been reported in the literature. Recently some groups are investigating possibilities to use high-level Petri nets as formal models for hardware/software co-design. Both groups of advanced applications are presented below.

The approach adopted at the University of Linköping (Sweden) is probably the most advanced in this area. It has been implemented in the CAMAD high-level synthesis system [71]. The system is specified in a sub-set of the VHDL language named S'VHDL [22]. Such specification is then mapped onto an extended timed Petri net (ETPN) model consisting of two separate but related parts, namely, a control structure and a data path [72]. The available parallelism is extracted and a register-transfer level (RTL) description of the hardware is produced [23]. The data path is represented as a directed graph where the nodes capture data manipulation units such as data storage elements and arithmetic operators, and the arcs represent the interconnection of these elements. The control structure is represented as a timed net with restricted firing rules, as defined in Peng's Ph. D. thesis [70]. Furthermore, the control structure communicates by issuing signals to and receiving conditional signals from the data path. On the one hand, the data path graph is an RTL description extracted directly from the VHDL specification. On the other hand, the timed Petri net description of the control structure as well as the corresponding communication signals are usually implemented as hardware through the synthesis of one or more finite state machines, depending on the style chosen by the designer of the system.

At Pièrre et Marie Curie University (France) [5] an approach was developed to the high level synthesis of embedded systems based on the description of the system in VHDL [11]. A model, named interpreted and timed Petri nets (ITPN) [10], has also been developed for the formal verification of the system's properties. A system of Boolean equations from the Petri net description is constructed first. System properties are then investigated using symbolic analysis of such equations [25]. Finally, the system is synthesized into an RTL VHDL version for implementation with ordinary synthesis tools for VLSI circuits [12].

There are several other approaches to translate VHDL specifications into different types of Petri nets for verification and synthesis purposes [20, 21, 22, 52, 54, 57, 58].

The other direction - from a Petri net to a synthesizable VHDL specification - has been also investigated. Synthesizing VHDL specifications requires writing according to restricted rules of the RTL. In particular the system must be clocked with a central clock signal. For that reason this method cannot be applied to asynchronous designs. On the other hand only controllers have been synthesized this way, as RTL specification of a data path would require scheduling and allocation of the resources, which has not been satisfactorily solved yet for Petri nets. The solutions for Petri net - based controller implementation through RTL VHDL have been reported [60, 73, 80, 83].

Despite several approaches to apply Petri nets to hardware-software co-design, there is no satisfactory solution for that yet. There are two main reasons: neither ordinary nor high-level nets have mechanisms of exception handling (necessary to represent reactive systems) and there is no widely accepted form of representing data in the nets. Most of the approaches tend to turn to object-oriented nets [27, 46, 47], but this approach needs more thorough theoretical investigation. Recently developed Place Chart Nets [39] is the first class of Petri Nets that has the notion of preemption and exception handling. Still more recent is the development of Petri Nets for Embedded Systems [53], which also seeks to fulfill the need for exception hadling mechanisms in Petri nets. In both cases, however, some details of the theory need to be further elaborated and a sound analysis environment is yet to be developed.

The work at the University of Minho (Portugal) is concentrated around a synchronous, hierarchical, object-oriented Petri net model [46, 47]. A set of analysis and synthesis tools has been developed there, aimed primarily at parallel controllers. Recently it is oriented towards co-design as well and targeted to special co-design architecture developed by the same group [47].

Object-oriented Petri nets are used also at Swiss Federal Institute of Technology [27]. The net there plays the role of a kernel language, representing different models of computation. A co-design system is being built around this model.

At the New University of Lisbon (Portugal) a framework is being developed for the design of complex reactive systems; applications to programmable controllers were developed. The system is defined as a reactive Petri net [32] which is a Petri net class based on colored Petri nets [38], synchronous interpreted Petri nets [28] and StateCharts [34]. The model includes transition priorities and hierarchical structure constructs. The analysis of the model is performed using the associated state space. The system is still under elaboration, however it is a goal of the development team to support the implementation of hardware on programmable logic components such as CPLDs and FPGAs, using direct and indirect synthesis of the Petri net model, within a co-design development environment.

A group at the University of Paderborn (Germany) is aiming at using "extended predicate/transition nets" to represent real-time embedded systems. The nets here are augmented with hierarchy and time and some of the first results are presented in [41].

The work developed at the University of Pernambuco (Brazil) restrict the application of Petri nets in co-design to some partitioning problems [48, 49]. They do not restrict themselves to one Petri net model only, but rather apply several nets that result from transformations. The nets used are hierarchical, interpreted and timed.

4. CONCLUSIONS

In this chapter we surveyed the results of the research on Petri net based synthesis of synchronous digital systems. We have restricted ourselves to synchronous domain as asynchronous design from Petri nets had been overviewed thoroughly elsewhere [84].

One important observation that can be drawn from this overview, and in particular from the bibliography below, is that apparently a lot of work in the field has been done. Moreover, the application domain of Petri nets in hardware design, previously restricted almost entirely to controllers, now is being extended to high- and system-level synthesis. This fact can be assigned to the availability of tools and methods for high-level Petri nets that allow representing more abstract concepts than ordinary Petri nets used in the past.

5. ACKNOWLEDGMENTS

J. Mirkowski's work has been done during a Visiting Research Fellowship at the University of Newcastle upon Tyne, UK, founded by NATO/Royal Society.

N.Marranghello's work has been done during a Post-Doctoral Research Fellowship at the Department of Computer Science of the University of Aarhus, Denmark, financed by Fundação de Amparo à Pesquisa do Estado de São Paulo. N.Marranghello wishes to thank Dr. Kurt Jensen, and the CPN group at Aarhus, for their support during the preparation of the paper.

The authors wish to thank Dr. Alex Yakovlev for several fruitful discussions during the preparation of the paper.

References

[1] M.Adamski (1987) *Direct Implementation of Petri Net Specification* 7th Int'l Conf. on Control Systems and Computer Science, pp.74-85.

[2] M.Adamski (1991) *Parallel Controller Implementation Using Standard PLD Software* In W.Moore and W.Luk (Eds.), Edited Papers from the Int'l Workshop on Field Programmable Logic and Applications, pp.296-304.

[3] A.Amroun and M.Bolton (1989) *Synthesis of Controllers from Petri Net Descriptions and Application of ELLA* Proc. of the IMEC-IFIP Int'l Workshop on Applied Formal Methods for Correct VLSI Design, pp.57-74, North-Holland.

[4] P.Ashar, S.Devadas and A.R.Newton (1990) *A Unified Approach to the Decomposition and Re-decomposition of Sequential Machines* Proc. of 27th ACM/IEEE DAC, pp.601-606.

[5] I.Augé, R.Bawa, P.Guerrier, A.Greiner, L.Jacomme and F.Pétrot (1997) *User Guided High Level Synthesis* Proc. of the IX IFIP Int'l Conf. on VLSI.

[6] M.Auguin, F.Boeri and C.Andre (1978) *New Design Using PLAs and Petri Nets* Int'l Symposium on Measurement and Control, pp.864-869, Athens.

[7] M.Auguin, F.Boeri and C.Andre (1980) *Systematic Method of Realization of Interpreted Petri Nets* Digital Processes, vol.6, pp.55-68.

[8] P.Azema, M.Diaz and R.Valette (1976) *Petri Nets as a Common Tool for Design Verification and Hardware Simulation* Proc. of 12th ACM/IEEE DAC, pp.109-116.

[9] M.Balakrishnan, A.K.Majumdar, D.K.Banerji and J.G.Linder (1988) Synthesis of Decentralised Controllers from High Level Description Microprocessing and Microprogramming, vol.22, pp.217-229, North-Holland.

[10] R.Bawa and E.Encrenaz (1996) *A Tool for Translation of VHDL Descriptions into a Formal Model and its Application to Formal Verification and Synthesis* LNCS, vol.1135, pp.471-474.

[11] R.Bawa and E.Encrenaz (1996) *A Platform for the Formal Verification of VHDL Programs* Proc. of the 4th Int'l Workshop on Symbolic Methods and Applications in Circuit Design, Louvain, Belgique.

[12] R.Bawa and L.Jancomme (1997) *Synthèse de Descriptions Comportementales Séquencielles en Conformité Avec la Sémantique VHDL* Actes de Colloque CAO de Circuits Intégrés et Systèmes, pp.303-306.

[13] K.Bilinski, M.Adamski, J.M.Saul and E.L.Dagless (1994) *Petri Net Based Algorithms for Parallel Controller Synthesis* IEE Proceedings, part E - Computers and Digital Techniques, 141(6)405-412, November.

[14] K.Bilinski, M.Adamski, J.M.Saul and E.L.Dagless (1994) *Parallel Controller Synthesis From a Petri Net Specification* Proc. of the European Design Automation Conf., pp.96-101, Grenoble.

[15] K.Bilinski, J.M.Saul and E.L.Dagless (1996) *Behavioural Synthesis of Complex Parallel Controllers* Proc. of the 9th Int. Conf. on VLSI Design, pp.186-191.

[16] B.Bruck, B.Kleinjohann, T.Kathofer and F.J.Ramming (1986) *Synthesis of Concurrent Modular Controllers from Algorithmic Description* Proc. of the 23rd ACM/IEEE DAC, pp.285-292.

[17] R.Bryant (1986) *Graph-based Algorithms for Boolean Function Manipulation* IEEE Trans. on Computers, 35(8)677-691.

[18] V.Catania, M.Malgeri and M.Russo (1997) *Applying Fuzzy Logic to Co-design Partitioning* IEEE Micro Magazine, 17(3)62-70.

[19] T.A.Chu (1987) *Synthesis of Self-timed VLSI Circuits from Graph-theoretic Specifications* Ph.D. Thesis, Mass. Inst. of Technol., MIT/LCS/TR-393, 189pp.

[20] G.Dohmen (1994) *Petri Net as Intermediate Representation Between VHDL and Symbolic Transition System* Proc. of the European DAC, pp.572-577.

[21] P.Eles, A.Doboli, Z.Peng and K.Kuchcinski (1995) *Timing Constraint Specification and Synthesis in Behavioral VHDL* Proc. of the European DAC.

[22] P.Eles, K.Kuchcinski, Z.Peng and M.Minea (1994) *Synthesis of VHDL Concurrent Processes* Proc. of the European DAC, pp.540-545.

[23] P.Eles, K.Kuchcinski and Z.Peng (1996) *Synthesis of Systems Specified as Interacting VHDL Processes* Integration - The VLSI Journal, 21(3)113-138.

[24] P.Eles, Z.Peng, K.Kuchcinski and A.Doboli (1996) *System Level Hardware/Software Partitioning Based on Simulation Annealing and Tabu Search* Journal of Design Automation for Embedded Systems, vol.2, pp.5-32.

[25] E.Encrenaz (1995) *A Symbolic Relation for a Subset of VHDL'87 Descriptions and its Application to Symbolic Model Checking* Proc. of the IFIP WG10.5 Advanced Research Working Conf. on Correct Hardware Design and Verification Methods, pp.328-342, LNCS, vol.987, Springer Verlag.

[26] J.Esparza, S.Römer and W.Vogler (1995) *An Improvement of McMillan's Unfolding Algorithm* Institut für Informatik, Technische Universität München, SFB-Bericht Nr.342/12/95-A, 16pp.

[27] R.Esser, J.W.Janneck and M.Naedele (1997) *Applying an Object-Oriented Petri Net Language to Heterogeneous System Design* Proc. of the Workshop on Petri Nets in System Engineering, Hamburg, Germany, http://www.informatik.uni-hamburg.de/TGI/aktuelles/pnse97/papers/janneck.ps.gz, 10pp., September.

[28] J.Fernandes, A.Pina and A.Proença (1995) *Simulação e Síntese de Controladores Paralelos Baseados em Redes de Petri* Anais do VII Simpósio Brasileiro de Arquiteturas de Computadores - Processamento Paralelo, pp.481-492.

[29] D.Gajski and R.Kuhn (1983) *Guests Editors' Introduction* IEEE Computer Magazine, 16(12)11-14.

[30] D.D.Gajski and L.Ramachandran (1994) *Introduction to High-Level Synthesis* IEEE Design & Test of Computers, 11(4)44-54.

[31] F.Garcia-Valles and J.M.Colom (1995) *A Boolean Approach to the State Machine Decomposition of Petri Nets with OBDD's* Proc. of IEEE Int'l Conf. on Systems, Man and Cybernetics.

[32] P.Goel (1981) *An Implicit Enumeration Algorithm to Generate Tests for Combinational Logic Circuits* IEEE Trans. on Computers, 30(3)215-222.

[33] L.Gomes and A.Garção (1985) *Programmable Controller Design Based on a Synchronized Colored Petri Net Model and Integrating Fuzzy Reasoning* Proc. of 16th Int'l Conf. on Appl. and Theory of Petri Nets, LNCS, vol.935, pp.218-237.

[34] D.Harel (1987) *Statecharts: A Visual Formalism for Complex Systems* Science of Computer Programming, 8.

[35] J.P.Hayes (1988) *Computer Architecture and Organization* McGraw-Hill Book Company (ISBN:0070273669).

[36] D.C.Hendry (1994) *Heterogeneous Petri Net Methodology for the Design of Complex Controllers* IEE Proceedings E - Comput. and Dig. Tech., 141(5)293-297.

[37] K.Jensen and G.Rozenberg (Eds.) (1991) *High-Level Petri Nets, Theory and Applications* Springer Verlag (ISBN:354054125X).

[38] K.Jensen (1997) *Coloured Petri Nets: Basic Concepts, Analysis Methods and Practical Use* Springer Verlag, vol.1 (ISBN:3540609431), 1992; vol.2 (ISBN:3540582762), 1994; and vol.3 (ISBN:3540628673), 1997.

[39] M. Kishinevsky, J. Cortadella, A. Kondratyev, L. Lavagno, A. Taubin, A. Yakovlev (1997) *Coupling Asynchrony and Interrupts. Place Chart Nets* Proc. of the 18th Int'l Conf. on Applications and Theory of Petri Nets, P. Azema and G. Balbo (Eds.), LNCS 1248, pp. 328-347, Springer-Verlag, Berlin.

[40] M.Kishinevsky, A.Kondratyev, A.Taubin and V.Varshavsky (1994) *Concurrent Hardware: The Theory and Practice of Self-timed Design* John Wiley & Sons Ltd. (ISBN:0471935360).

[41] B.Kleinjohann, J.Tacken and C.Tahedl (1997) *Towards a Complete Design Method for Embedded Systems Using Predicate/Transition Nets* Proc. of the Int'l Conf. on Computer Hardware Design Languages, Toledo, Spain, March.

[42] A.Kondratyev, M.Kishinevski, B.Lin, P.Vanbekbergen and A.Yakovlev (1994) *Basic Gate Implementation of Speed-independent Circuits* Proc. of the 31st ACM/IEEE DAC, pp.56-62.

[43] T.Kozlowski (1993) *Petri Net Based CAD Tools for Parallel Controller Synthesis* PhD thesis, University of Bristol.

[44] T.Kozlowski, E.Dagless, J.Saul, M.Adamski and J.Szajna (1995) *Parallel Controller Synthesis Using Petri Nets* IEE Proceedings, part E - Computers and Digital Techniques, 142(4)263-271.

[45] K.C.Leung, C.Michel and P.Le Beux (1977) *Logical Systems Design Using PLAs and Petri Nets - Programmable Hardware Systems* B.Gilchrist (Ed.), IFIP Information Processing 77, pp.607-611.

[46] R.Machado, J.Fernandes and A.Proença (1997) *Specification of Industrial Digital Controllers with Object-Oriented Petri Nets* IEEE Int'l Symposium on Industrial Electronics, Guimarães, Portugal.

[47] R.Machado, J.Fernandes and A.Proença (1998) *An Object-Oriented Model for Rapid Prototyping of Data Path/Control Systems - A Case Study* Proc. of the 9th IFAC Symposium on Information Control in Manufacturing, Nancy, France.

[48] P.Maciel, E.Barros and W.Rosenstiel (1997) *Using Petri Nets to Compute Communication Cost for Hardware/Software Codesign* Proc. of the Int'l Workshop on Rapid System Prototyping, Leuven, Belgium, June.

[49] P.Maciel, T.Maciel, E.Barros and W.Rosenstiel (1998) *A Petri Net Approach to Compute Load Balance in Hardware/Software Codesign* Proc. of the Conf. on High Performance Computing, Boston, Massachusetts, U.S.A.

[50] K.McMillan (1995) *Trace Theoretic Verification of Asynchronous Circuits Using Unfoldings* Proc. of the 7th Int'l Conf. on Computer Aided Verification, pp.180-194, P.Wolper (Ed.), Springer Verlag.

[51] G.de Micheli (Editor) (1997) *Special Issue on Hardware/Software Co-design* IEEE Proceedings, vol.85, no.3.

[52] J.Mirkowski, K.Bilinski and E.L.Dagless (1996) *Petri Net Modeling of VHDL Simulation Cycle for High Level Synthesis Purposes* Proc. of the SIG-VHDL Spring'96 Working Conf. VHDL-Forum Europe, pp.35-46, Dresden, Germany.

[53] J.Mirkowski and A.Yakovlev (1998) *A Petri Net Model for Embedded Systems Design and Diagnostics of Circuits and Systems* Workshop, Szczyrk, Poland, September 2-4.

[54] J.Mueller and H.Kraemer (1993) *Analysis of Multi-Process VHDL Specifications with a Petri Net Model* Proc. of the European DAC, pp.474-479, September.

[55] T.Murata (1989) *Petri Nets: Properties, Analysis and Applications* Proceedings of the IEEE, 77(4)541-580.

[56] R.Nishimura and S.G.Zaky (1989) *Synthesis of a Petri Net Based Control Flow Model* Procceddings of the IEEE Int'l Symp. on Ccts. and Sys., vol.1, pp.313-318.

[57] S.Olcoz and J.M.Colom (1993) *Towards a Formal Semantics of IEEE Std. VHDL 1076* Proc. of the European DAC, pp.526-531, Hamburg, Germany.

[58] S.Olcoz and J.M.Colom (1995) *A Colored Petri Net Model of VHDL* Formal Methods in System Design, vol.7, pp.101-123.

[59] J.V.Oldfield and R.C.Dorf (1995) *Field Programmable Gate Arrays* Wiley-Interscience Publication (ISBN:0471556653).

[60] J.Pardey (1993) *Parallel Controller Synthesis for VLSI Applications* PhD thesis, University of Bristol, UK.

[61] J.Pardey, A.Amroun, M.Bolton and M.Adamski (1994) *Parallel Controller Synthesis for Programmable Logic Devices* Microprocessing and Microprogramming, vol.18, pp.451-457, North-Holland.

[62] J.Pardey and M.Bolton (1991) *Logic Synthesis of Synchronous Parallel Controllers* Proc. of the IEEE Int'l Conf. on Computer Design, pp.454-457.

[63] J.Pardey and M.Bolton (1992) *Parallel Controller Synthesis for Concurrent Data Paths* IFIP Workshop on Control Dominated Synthesis from a Register Transfer Level Description, pp.16-19.

[64] J.Pardey, T.Kozlowski, J.Saul and M.Bolton (1992) *State Assignment Algorithms for Parallel Controller Synthesis* Proc. of the IEEE Int'l Conf. on Computer Design, pp.316-319.

[65] M.R.K.Patel (1990) *Random Logic Implementation of Extended Timed Petri Nets* Microprocessing and Microprogramming, 30(1-5)313-320.

[66] S.Patil (1972) *Circuit Implementation of Petri Nets Computation Structures* Group Memo 73, Project MAC, Mass. Inst. of Technol., USA, 15pp., December.

[67] S.Patil (1974) *Cellular Arrays for Asynchronous Control* Conf. Record of the 7th Annual Workshop on Microprogramming, pp.178-185, Palo Alto, CA, U.S.A.

[68] Z.Peng (1986) *Synthesis of VLSI Systems with the CAMAD Design Aid* Proc. of the 23rd ACM/IEEE DAC, pp.278-284.

[69] Z.Peng and K.Kuchcinski (1986) *Synthesis of Control Structures from Petri Net Descriptions* Microprocessing and Microprogramming, vol.18, pp.335-340.

[70] Z.Peng (1987) *A Formal Methodology for Automated Synthesis of VLSI Systems* Ph.D. Thesis, nr.170, Department of Computer and Information Sciences, Linköping University, Sweden.

[71] Z.Peng and K.Kuchcinski (1994) *Automated Transformation of Algorithms into Register Transfer Level Implementation* IEEE Trans. on Computer Aided Design of Integrated Circuits and Systems, 13(2)150-166.

[72] Z.Peng and A.Törne (1993) *A Petri Net Based Modelling and Synthesis Technique for Real-time Systems* Proc. of the 5th Euro-Micro Workshop on Real Time Systems.

[73] R.Rao, G.Swaminathan, B.W.Johnson and J.H.Aylor (1994) *Synthesis of Reliability Models from Behavioral-Performance Models* Proc. of the Reliability and Maintainability Symposium, pp.292-298.

[74] W.Reisig (1992) *A Primer in Petri Net Design* Springer Verlag (ISBN:3540520449).

[75] J.Rozenblit and K.Buchenrieder (Eds.) (1995) *Co-design: Computer-aided Software/Hardware Engineering* IEEE Press, (ISBN:0780310497).

[76] M.Sawasaki, C.Y.-Couvreur and B.Lin (1995) *Externally Hazard-free Implementation of Asynchronous Circuits* Proc. of the 32nd ACM/IEEE DAC, pp.718-724, San Francisco, CA, U.S.A.

[77] A.Semenov and A.Yakovlev (1996) *Verification of Asynchronous Circuits Using Time Petri Net Unfolding* Proc. of the 33rd ACM/IEEE DAC, pp.59-62.

[78] J.Stewart, E.Dagless, D.Milford and O.Miles (1991) *A Petri Net Based Framstore* Proc. of the Int'l Workshop on Field Programmable Logic and Applications, pp.332-342.

[79] E.Stoy and Z.Peng (1997) *Inter-domain Movement of Functionality as a Repartitioning Strategy for Hardware/Software Co-design* Journal of Systems Architecture, vol.43, pp.87-98.

[80] G.Swaminathan, R.Rao, J.H.Aylor and B.W.Johnson (1994) *A VHDL Based Enviroment for System Level Design and Analysis* Proc. of the Reliability and Maintainability Symposium, pp.110-116.

[81] J.M.Toulotte and J.P.Parsy (1979) *A Method for Decomposing Interpreted Petri Nets and its Utilization* Digital Processes, vol.5, pp.223-234.

[82] A.Valmari (1991) *Stubborn Attack on State Explosion* Formal Methods in System Design 1(1)297-322.

[83] W.Wegrzyn, M.Adamski and J.Monteiro (1996) *VHDL Simulation of Xilinx-FPGA-based Concurrent Controller* Proc. of a Workshop on Application of Programmable Logic, pp.12-15, Lisbon, Portugal.

[84] A.Yakovlev and A.Koelmans (1998) *Petri Nets and Digital Hardware Design* Advances in Petri Nets, LNCS, Springer Verlag.

[85] A.Yakovlev, A.Koelmans, A.Semenov and D.Kinniment (1996) *Modelling, Analysis and Synthesis of Asynchronous Control Circuits Using Petri Nets* Integration - The VLSI Journal, 21(3)143-170.

Chapter 8

DERIVING SIGNAL TRANSITION GRAPHS FROM BEHAVIORAL VERILOG HDL

Ivan Blunno
Politecnico di Torino
Torino, Italy
blunno@gandalf.polito.it

Luciano Lavagno
Universitá di Udine
Udine, Italy
lavagno@uniud.it

Abstract   We propose a design flow for asynchronous circuits that closely mimics the standard synchronous ASIC design flow. Key elements of the flow are HDL-based specification, logic synthesis and physical design. In this work we present a proposal for using a standard HDL, Verilog, to specify an asynchronous *control* circuit at the behavioral level. This pecification is automatically translated in a Signal Transition Graph, that can then be automatically synthesized by existing tools.

Advantages of this methodology include rapid path to implementation, re-use of simulation patterns between pre- and post-synthesis steps, and designer familiarity with the specification language.

Keywords: asynchronous control circuits, HDL, Petri nets, Signal Transition Graphs, Verilog, VHDL

1.   INTRODUCTION

Language-based design has become the standard in synchronous circuit design flows. The designer specifies the circuit functionality by using a Hardware Description Language (most often restricted to a synthesizable subset), simulates it to verify its correctness within an environment scenario (testbench), synthesizes it to a netlist of standard gates

A. Yakovlev et al.(eds.), Hardware Design and Petri Nets, 151-170
© 2000 *Kluwer Academic Publishers.*

(standard cells, gate arrays, field programmable gate arrays), places and routes it, and then finally re-simulates everything with the extracted back-annotated delays.

This flow, which can incorporate more complex library components such as data path generators, RAMs, register files, and so on, is based on *inference rules* that allow one to use various HDL constructs to specify the circuit functionality at an *abstract*, extended Finite State Machine level that is amenable to automated synthesis.

The back-end is based again on an abstraction, that separates *functionality*, as embodied by the gate netlist, from *timing*, as extracted from the cell library and the interconnect layout.

The purpose of this work is to investigate if and how this flow, that is now the reference point for modern *synchronous* design, is applicable to *asynchronous* circuits as well. This decision was motivated by the emergence of several logic synthesis tools for asynchronous circuits (e.g., [7]), and by the common assumption that physical design tools (place and route, extract, ...) can be made to work also for asynchronous circuits.

This assumption is justified from the theoretical standpoint, because the most common design styles for asynchronous circuits (e.g., Speed-Independent circuits [12, 8], Delay-Insensitive circuits [11, 18, 20], and Burst-Mode Finite State Machines [13, 23]) use, up to a point, the very same *separation between functionality and timing* that is the basis of synchronous design, both at the logical and at the physical level. Asynchronous design methodologies achieve this goal by making *timing assumptions* that are roughly equivalent to the synchronous assumptions that setup and hold times are satisfied, that the clock skew is limited, and so on. Asynchronous timing assumptions (e.g., isochronic forks for SI circuits, delay bounds within modules for DI circuits, or Fundamental Mode for BM circuits) are much more local and distributed in nature than their synchronous counterparts. This means that asynchronous assumptions are harder to satisfy *today* (since placement and routing algorithms have problems satisfying a large number of constraints), but that they can potentially *scale* better than synchronous constraints as feature size decreases and chip size increases. In fact, physical design tools for Deep-Sub-Micron technologies already have to manage an increasingly large number of local constraints, due to the increasing importance of wire delays over gate delays[1].

---

[1] We do not want to delve here into the long-standing debate about the relative scalability of the SI, DI or FM timing assumptions.

In practice, standard cell libraries, and even Field-Programmable Gate Arrays, have been successfully used in several asynchronous designs [19, 2], thus confirming the validity of our assumption.

So far, however, the design of asynchronous circuits has been based on "ad hoc" languages, such as Tangram [20], or variants of Communicating Sequential Processes [10], or graphical waveform editors [1]. Some tools (e.g., the above mentioned Petrify, SIS from U.C. Berkeley and Assassin from IMEC [22]) use as specification an interpreted form of Petri nets called Signal Transition Graph [15, 4], that is represented using an internal file format that is unreadable by the average designer.

"Ad hoc" asynchronous languages have a few advantages over standard ones. In particular, they are better suited to model asynchronous circuits, while, as we will see later, languages from the synchronous domain require one to potentially restrict the set of specifiable asynchronous circuits to those that can be expressed within a syntax and semantics originally conceived for a different application. For example, Signal Transition Graphs can be considered as a formal representation of timing diagrams, a specification mechanism that is quite familiar to designers. Moreover, languages such as Tangram and CSP make it easier to specify only "meaningful" asynchronous circuits (e.g., circuits that are Delay Insensitive by construction in the case of Tangram).

In this work we chose to use a *standard language*, Verilog HDL [17], because:

- it is well-known to designers, and hence it reduces the amount of re-training and the potential resistance to change,

- it has a wealth of supporting tools, such as editors, translators to and from other formats, simulators, physical design systems, timing analyzers[2], and so on,

- it potentially enables a *mixed* synchronous/asynchronous design flow well suited, for example, to mostly synchronous circuits in which asynchronous components (e.g., interfaces) would currently require to be designed by hand.

None of these advantages is directly accessible to designers using non-standard languages such as timing diagrams, Tangram and so on, since they imply to enter the classical design flow only as a *back-end* activity, after synthesis (e.g., as a standard cell netlist written in Verilog). This has all the problems of a mixed-language design flow (need to translate

---

[2]With some care, of course, due to the cyclic nature of asynchronous circuits.

test-benches, need to learn multiple languages, ...), and thus we are proposing an alternative to it.

This is not the first work attempting to specify asynchronous circuits using a standard language. In particular, Vanbekbergen et al. proposed in [21] to use VHDL, with overall goals that were very similar to ours[3], and translating it into Signal Transition Graphs as in our case. However, they forced the designer to explicitly (and unnaturally) list all places and transitions of the underlying Petri net. Our work, in contrast, attempts to use *only natural behavioral constructs* when specifying asynchronous control circuits, and thus should be easier to use, as our examples will show.

This paper is *not* a theoretical contribution, and does not present exciting new experimental results (even though we hope that our tool will be used for real, exciting designs). It is a *methodology* paper, aimed at showing how non-specialists can be introduced to asynchronous design via an appropriate choice of tools and languages. It proposes an *Asynchronous Synthesizable Subset of Verilog* that seems sufficient to specify at least *control* circuits. An extension to handle also data paths is under way.

The paper is organized as follows. In Section 2. we describe our input and output formalisms and the design flow. In Section 3. we describe the subset of Verilog that we support. In Section 4. we provide some examples demonstrating the flexibility and readability of our modeling style. In Section 5. we draw some conclusions and outline many opportunities for future work.

2.   VERILOG-BASED DESIGN OF ASYNCHRONOUS CIRCUITS

The specification language: Verilog.   Our choice of specification language was Verilog. Verilog has the advantage over VHDL that it already allows one, by means of the fork-join construct, to mix more or less arbitrarily concurrency with sequencing and data-dependent choice. The only constraint (that in fact can be a limitation when one would like to exploit at best the potential of asynchronous specifications operating in Input/Output mode such as Signal Transition Graphs) is to combine sequencing, choice and concurrency in a *structured*, well-nested fashion due to the "structured programming" language style [17]. Verilog also already includes an "uninterpreted event" construct that is more high-

---

[3]Synopsys at that time seemed interested in selling an asynchronous companion/extension to their Design Compiler^{TM}.

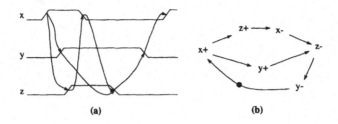

Figure 8.1   Example of a Signal Transition Graph

level than VHDL's strictly value-based semantics, and that could be used in the future, for example, to specify handshakes for which the encoding still has to be specified ([3]).

VHDL, that would have been a natural choice considering the several existing examples of formal semantics based on Petri nets, on the other hand offers only two levels of concurrency (among processes) and sequencing/choice (within a process). This is too limiting for asynchronous circuits, since the fine interplay between concurrency, sequencing and choice is an essential advantage of asynchronicity versus synchronicity. For this reason [21] was forced to explicitly list Petri net transitions and places, instead of using VHDL behavioral constructs, and thus had to compromise ease of use.

The output language: Signal Transition Graphs.   The output language that we chose for this project is Signal Transition Graphs (STGs). STGs [15, 4] are interpreted Petri nets that formalize the notion of Timing Diagrams (see an example of the correspondence in Figure 8.1) and that can be synthesized into asynchronous circuits [7]. In particular, the Petri Nets that are generated by our tool satisfy the Free Choice restriction [6], and are thus amenable to very efficient structure-based synthesis methods that do not require the expoenential Petri net reachability analysis step [14].

The design flow.   The design flow that we advocate is shown in Figure 8.2. In particular, in this paper we focus on the only phase that is currently missing: the translation from a Verilog specification to an STG in a format suitable for state-of-the-art asynchronous synthesis tools. The opposite step, from the output format, e.g., of Petrify to Verilog or VHDL, is already implemented with simple converters that can be found, e.g., at the same URL where we distribute our tool (see Section 5.).

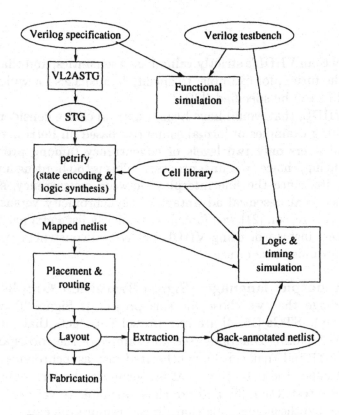

Figure 8.2   Language-based asynchronous design flow

State encoding and logic synthesis can be performed, for example, with Petrify [7], while physical design, extraction and netlist comparison, as argued above, can be tackled with standard tools.

Simulation, of course, can also be performed with a standard (commercial) Verilog simulator at all levels (specification, logic gates, extracted netlist), thus enabling to one to use the same *testbench* for:

- functional validation,

- logic simulation and performance analysis,

- simulation with extracted parasitics and verification of timing assumptions (e.g., of *isochronic forks*).

3. THE ASYNCHRONOUS SYNTHESIZABLE SUBSET OF VERILOG

The subset of Verilog that can be synthesized by the current version of our tool (called *vl2astg*) includes the following constructs:

1. module, input, output and reg[4] declarations. Signals must be single-bit (as mentioned above, multi-bit signals and data paths will be handled in the future).

   Only one module per STG is currently supported. Hierarchical specifications are of course possible by using a structured methodology in which non-leaf modules can only instantiate other modules and have no behavior of their own. The hierarchy would then remain exactly the same after synthesis, but instantiate the gate-level netlist of the leaf modules produced by logic synthesis, instead of the behavioral specification that is input to *vl2astg*.

   For example, the following fragment:

   ```
   module example(a, b, c, d);
     input a, b;
     output c, d;
     reg e, f, g;
   ```

   declares a module with two inputs, two outputs and three internal signals.

2. initial, to specify the initialization sequence. For example, the following fragment:

---

[4]Despite the name, a reg denotes an internal signal of a module, and hence of the STG.

```
initial
begin
    c = 1;
    d = 1;
    e = 0;
    f = 1;
    g = 0;
end
```

is a simple initialization, consisting of a set of *sequentially ordered* (due to the use of the **begin** ... **end** construct described more in detail below) assignments.

3. **always**, to specify the main operational cycle of the controller. For example, the following fragment (in Verilog | denotes *or*, & denotes *and*, and ! denotes *not*):

```
always
begin
    wait(a);
    c = 0;
    wait (!a);
    c = 1;
end
```

specifies the operation of an *inverter*, with input **a** and output **c**, used by an environment that waits for the output to change before the input can be changed.

4. Assignment of 1 or 0 to an output variable. Note that assignment is synthesized into an STG following the Verilog assignment semantics. This means that

```
    c = 1;
```

does not necessarily result in a rising edge of signal c (c^+ in STG notation), but rather in an STG fragment that tests for the value of c in the current state, and assigns 1 to it only if the current value is 0, as shown in Figure 8.3. The top place joins this fragment with those causally preceding it, the bottom place joins it with those causally following, and the two places labeled $c = 0$ and $c = 1$ (updated also by other rising and falling transitions of c, of course) keep track of the current value of c. ϵ is a "dummy" STG transition, that does not change the value of any signal.

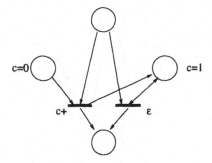

Figure 8.3 STG fragment for signal assignment

We chose to implement assignment in this slightly complex way because it is consistent with the simulation semantics, and thus does not cause any mismatch between the behavior of the simulator and that of the generated STG. Of course, care must be exercised to make sure that the resulting STG can be implemented as a Speed-Independent circuit [8][5].

5. **wait** to specify waiting for an input *level*. This statement generates a fragment of STG like that for assignment, rather than waiting for a *transition* of the input signal. Thus the following fragment:

    ```
    wait(!a);
    ```

    does not just generate a single a^- STG transition in general, but a subgraph like that shown in Figure 8.4 (and similar to that of Figure 8.3).

    A statement waiting for input transitions, thus corresponding directly to an STG transition, will be added in the near future, using the @ **posedge** and @ **negedge** Verilog syntax.

6. **begin-end** blocks to specify sequencing. The following fragment:

    ```
    begin
        d = 0;
        wait (! b);
        d = 1;
    ```

---

[5]Petrify currently provides this information. We have not yet devised any mechanism, however, to back-annotate error messages from Petrify to the Verilog file level, in order to ensure a more user-friendly usage.

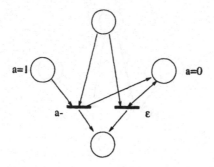

Figure 8.4   STG fragment for signal level waiting

```
    wait (! a);
  end
```

denotes (assuming, for the sake of simplicity, that the initial value of signals is a=1, b=1, d=1) the *sequence* of STG transitions d^-, b^-, d^+, a^-.

7. **fork-join** blocks to specify concurrency. For example, the following fragment:

```
fork
  c = 1;

  begin
    d = 0;
    wait (! b);
    d = 1;
    wait (! a);
  end
join
```

denotes (assuming, for the sake of simplicity, that the initial value of signals is a=1, b=1, c=0, d=1) the *concurrent* execution of c^+ and of the sequence d^-, b^-, d^+, a^- in the STG.

8. **if** (the **else** clause is currently not allowed, for reasons discussed below), to specify non-deterministic input choice (i.e., a branch in the execution flow depending on a Boolean condition).

The specification of input choice in Verilog is actually the least intuitive part of our STG specification style, because

- in the STG, it is denoted by a place with more than one successor, and hence it denotes *waiting for the first (and only) one among a set of transitions to occur*.

- in Verilog this waiting followed by a choice must be modeled by using two statements, as in the following sample fragment:

```
wait (! a | b) begin
  if (! a) c = 1;

  if (b) begin
  ...
```

The first `wait` specifies the waiting for *one* of a disjunction of possible input transitions. Note that the environment of the circuit must *guarantee that these transitions are mutually exclusive*, since there is no easy way to model the general, simulation-oriented and OR-causal `wait` statement with STGs. Then we either assign 1 to c if a^- has occurred, or execute some (unspecified) sequence of transitions if b^+ has occurred. Thus the set of conditions tested in the `if` statements[6] must be the same as those *OR*ed together in the `wait` statement.

A *select*-style statement in Verilog, modeling wait and branch together, would obviously make the specification of choice more natural. It is unfortunate that the language designers did not include one, but their goal was clearly never to design a synthesis-oriented language (much less an asynchronous synthesis-oriented one).

The constructs summarized above have been chosen to *minimally* support *in a behavioral specification style* the main paradigms of asynchronous control: *sequencing*, *concurrency* and *non-deterministic choice*. All these constructs are naturally mapped into STG patterns, and "stitched" together appropriately by *vl2astg*.

For example, consider the more complete Verilog specification shown in Figure 8.5. It shows a typical mix of sequencing, concurrency and choice. First of all, the asynchronous circuit specified by this Verilog module initializes both outputs to 1 (as specified by the `initial` construct). Then it must wait for a to rise (this implicitly means that the

---

[6]Only single signals, in positive or negative form, are currently supported. The extension allowing two-level and/or expressions is left to future work.

environment will keep a low initially). After that, it must lower c and wait for

- either a to fall, in which case it must raise c again,

- or b to rise, in which case it concurrently

  - raises c,

  - completes a sequential "handshake" between d, b and a.

Once one cycle has completed, with c at 1 and a at 0, the next cycle can begin.

The STG can be implemented as an asynchronous circuit by *petrify* [7], producing the following equations (csc0 is a state signal, added by *petrify* to make the circuit implementable):

```
c = csc0 | !a;
d = !b;
csc0 = a csc0 | b;
```

The resulting STG is shown in Figure 8.6 *after simplification by Petrify* (that can also manipulate and minimize Signal Transition Graphs [5]). In this case, assignments always result in output signal transitions, and waits always result in input signal transitions, and the STG can be simplified considerably with respect to the version produced by *vl2astg* by fragment stitching.

It should be obvious that a simulatable Verilog specification is much easier to write and maintain than the corresponding STG, due to the reasons listed in Section 1.. Just as one example, Verilog forces sequencing to be expressed by syntactic vicinity. In the STG format used by Petrify and other tools, places modeling sequencing can be put in freeformat, and one in general has to scan the whole file in order to find the predecessors and successors of a given transition. In fact, STGs were born for *visualization*, not textual editing, yet no graphical STG editor (just a pretty printing tool) is available as of today.

4.   EXAMPLES OF ASYNCHRONOUS VERILOG STG SPECIFICATIONS

In this section we introduce a more complex specification example, a VME bus master interface from [9].

Figure 8.7 shows the original timing diagram, from the official VME specification. Figures 8.8, 8.9 and 8.10 show the Verilog specification for this timing diagram. The resulting STG, with 94 transitions and

```
module example(a, b, c, d);
  input a, b;
  output c, d;

initial
begin
  c = 1;
  d = 1;
end

always
begin
  wait(a);
  c = 0;
  wait (! a | b) begin
    if (! a) c = 1;

    if (b) begin
    fork
      c = 1;

      begin
        d = 0;
        wait (! b);
        d = 1;
        wait (! a);
      end
    join
    end
  end
end
endmodule
```

Figure 8.5   Example of Verilog specification

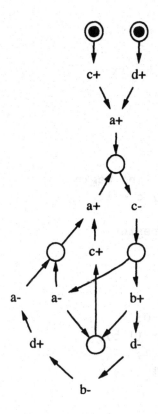

Figure 8.6   STG synthesized from Figure 8.5

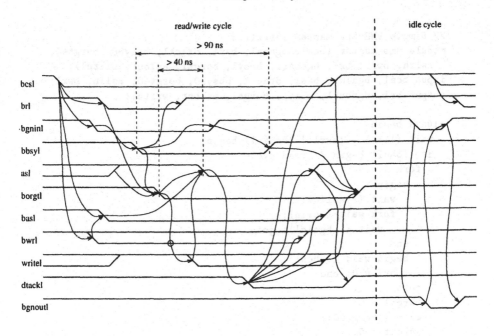

Figure 8.7 The timing diagram of an asynchronous VME bus master interface

123 places is too complex to be drawn in a readable figure. It was also successfully synthesized by Petrify into a Speed-Independent circuit.

This example shows a complex interplay of concurrency and sequencing (while the amount of choice is fairly limited). It also shows some of the limitations of the Verilog "structured" programming style. For example, consider the behavior of the rising edges of *bwrl, bcsl* and *borgtl* in Figure 8.7. The first two should both be caused in parallel by the falling edge of *dtackl*, and *bcsl* should cause the rising edge of *borgtl*. First of all, we had to add some links that are missing in the timing diagram, e.g., that from $bwrl^+$ to some other transition, in order to keep the specification Speed-Independent and consistent. Moreover, this behavior would be combined in a non-series/parallel fashion with the other rising edges that close the write cycle. For this reason, the STG of Figure 8.9 is forced to specify that *bcsl+* causes *borgtl+* and that causes $bwrl^+$. This is a less concurrent specification, possibly leading to a slower implementation.

5.   CONCLUSION

We have described a first step towards the implementation of a language-based design flow for asynchronous circuits. The flow is based on a standard language, Verilog, and supports:

```
// Simple VME bus master interface
module vme_master (bcsl, bgninl, basl, dtackl, bbsy90, borgt40,
   aslin, bwrl, brl, bgnoutl, bbsyl, borgtl, aslout, writel);
input bcsl, bgninl, basl, dtackl, bbsy90, borgt40, aslin, bwrl;
output brl, bgnoutl, bbsyl, borgtl, aslout, writel;

always begin
  wait (!basl | !bwrl | !bgninl) begin
  if (!basl) begin // read cycle
    fork
      begin
        wait (! aslin);
        fork wait (aslin);
          wait (! bgninl); join
      end
      begin wait (! bcsl);
        brl = 0; end
    join
    borgtl = 0;
    wait (!borgt40);
    aslout = 0;
    fork
      begin
        bbsyl = 0;
        fork
          begin brl = 1;
            wait (bgninl); end
          wait (!bbsy90);
        join
        bbsyl = 1;
      end
      begin
        wait (!aslin); wait (!dtackl);
        fork
          begin aslout = 1;
            wait (aslin); end
          wait (bcsl);
        join
      end
    join
    borgtl = 1;
    fork wait (bbsy90); wait (borgt40);
      wait (dtackl); wait (basl); join
  end
  ...
```

Figure 8.8   The Verilog specification of the VME read cycle

```
...
if (!bwrl) begin // write cycle
  fork
    wait (!basl);
    begin
    fork
      begin wait (!bcsl);
        brl = 0; end
      begin
        wait (!aslin);
        fork wait (!bgninl);
          wait (aslin); join
      end
    join
    end
    borgtl = 0;
    wait (!borgt40);
    writel = 0;
  join
  aslout = 0;
  fork
    begin
      bbsyl = 0;
      fork
        begin brl = 1;
          wait (bgninl); end
        wait (!bbsy90);
      join
      bbsyl = 1;
    end
    begin
      wait (!aslin);
      wait (!dtackl);
      fork
        begin aslout = 1;
          wait (aslin); end
        writel = 1;
        wait (bcsl);
      join
    end
  join
  borgtl = 1;
  fork wait (basl); wait (bwrl);
    wait (dtackl); wait (borgt40);
    wait (bbsy90); join
end
...
```

Figure 8.9 The Verilog specification of the VME write cycle

```
  ...
  if (!bgninl) begin
  // pass daisy chain
    fork
      begin bgnoutl = 0; wait (bgninl);
        bgnoutl = 1; end
      wait (!bcsl);
    join
    wait (bcsl);
  end
  end
end
endmodule
```

Figure 8.10   The Verilog specification of the VME idle cycle

- functional (pre-synthesis) simulation,

- synthesis using existing research tools for asynchronous circuits,

- post-synthesis simulation and layout using commercial tools originally developed for synchronous circuits.

In the future, we are planning to extend the tool to handle the *data path* as well, thus providing a level of abstraction comparable to Register Transfer Level in the synchronous case. We are planning to support an *automated* micropipeline-style decomposition, with a synchronous data path (synthesized with standard tools) and an asynchronous control unit using 4-phase handshakes. As usual, it will be crucial to automate as many design steps as possible. In particular, we are planning to use synchronous timing analysis in the data path to automatically generate the matched delay lines required to satisfy the data bundling constraints [16].

The tool described in this paper is publicly available at URL

http://polimage.polito.it/~blunno/vl2astg/

Feedback from potential users is of course encouraged and thanked in advance.

Acknowledgments

We would like to thank the members of the ACiD Esprit Working Group for their participation in a discussion on language-based design of asynchronous circuits in January, 1998. A summary of the discussion, that partially inspired us for this work, can be found at http://polimage.polito.it/acid

References

[1] Borriello, G. (1988). *A New Interface Specification Methodology and its Application to Transducer Synthesis*. PhD thesis, U.C. Berkeley. (technical report UCB/CSD 88/430).

[2] Brunvand, E. and Sproull, R. F. (1989). Translating concurrent programs into delay-insensitive circuits. In *Proceedings of the International Conference on Computer-Aided Design*, pages 262–265.

[3] Burns, S. and Martin, A. (1987). A synthesis method for self-timed VLSI circuits. In *Proceedings of the International Conference on Computer Design*.

[4] Chu, T.-A. (1986). On the models for designing VLSI asynchronous digital systems. *Integration: the VLSI journal*, 4:99–113.

[5] Cortadella, J., Kishinevsky, M., Lavagno, L., and Yakovlev, A. (1998). Deriving petri nets from finite transition systems. *IEEE Transactions on Computers*, 47(8):859–882.

[6] Desel, J. and Esparza, J. (1995). *Free choice Petri nets*. Cambridge University Press, New York.

[7] J. Cortadella et al. (1997). UPC/DAC VLSI CAD Group: Petrify. See
http://www.ac.upc.es/~vlsi/petrify/petrify.html.

[8] Kondratyev, A., Kishinevsky, M., Lin, B., Vanbekbergen, P., and Yakovlev, A. (1994). Basic gate implementation of speed-independent circuits. In *Proceedings of the Design Automation Conference*.

[9] Lavagno, L. and Sangiovanni-Vincentelli, A. (1993). *Algorithms for synthesis and testing of asynchronous circuits*. Kluwer Academic Publishers.

[10] Martin, A. (1990). Programming in VLSI: From communicating processes to delay-insensitive circuits. In Hoare, C. A. R., editor, *Developments in Concurrency and Communications*, The UT Year of Programming Series. Addison-Wesley.

[11] Molnar, C. E., Fang, T.-P., and Rosenberger, F. U. (1985). Synthesis of delay-insensitive modules. In *Chapel Hill Conference on VLSI*, pages 67–86.

[12] Muller, D. E. and Bartky, W. C. (1959). A theory of asynchronous circuits. In *Annals of Computing Laboratory of Harvard University*, pages 204–243.

[13] Nowick, S. M. and Dill, D. L. (1991). Automatic synthesis of locally-clocked asynchronous state machines. In *Proceedings of the International Conference on Computer-Aided Design*.

[14] Pastor, E. and Cortadella, J. (1993). Polynomial algorithms for the synthesis of hazard-free circuits from signal transition graphs. In *Proceedings of the International Conference on Computer-Aided Design*.

[15] Rosenblum, L. Y. and Yakovlev, A. V. (1985). Signal graphs: from self-timed to timed ones. In *International Workshop on Timed Petri Nets, Torino, Italy*.

[16] Sutherland, I. E. (1989). Micropipelines. *Communications of the ACM*. Turing Award Lecture.

[17] Thomas, D. and Moorby, P. (1995). *The Verilog hardware description language*. Kluwer Academic.

[18] Udding, J. T. (1986). A formal model for defining and classifying delay-insensitive circuits and systems. *Distributed Computing*, 1:197–204.

[19] van Berkel, K., Burgess, R., Kessels, J., Peeters, A., et al. (1995). A single-rail re-implementation of a DCC error detector using a generic standard-cell library. In *Proc. Second Working Conf. on Asynchronous Design Methodologies, London*.

[20] van Berkel, K., Kessels, J., Roncken, M., Saejis, R., and Schalij, F. (1991). The VLSI-programming language Tangram and its translation into handshake circuits. In *Proceedings of the European Design Automation Conference (EDAC)*, pages 384–389.

[21] Vanbekbergen, P., Wang, A., and Keutzer, K. (1995). A design and validation system for asynchronous circuits. In *Proceedings of the Design Automation Conference*.

[22] Ykman-Couvreur, C., Lin, B., Goossens, G., and Man, H. D. (1993). Synthesis and optimization of asynchronous controllers based on extended lock graph theory. In *Proceedings of the European Design Automation Conference (EDAC)*.

[23] Yun, K. Y. and Dill, D. L. (1992). Automatic synthesis of 3D asynchronous state machines. In *Proceedings of the International Conference on Computer-Aided Design*.

Chapter 9

THE DESIGN OF THE CONTROL CIRCUITS FOR AN ASYNCHRONOUS INSTRUCTION PREFETCH UNIT USING SIGNAL TRANSITION GRAPHS

An asynchronous instruction prefetch unit design

Suck-Heui Chung and Steve Furber
Department of Computer Science
University of Manchester
Oxford Road, Manchester
M13 9PL, United Kingdom
{chung,sfurber}@cs.man.ac.uk

Abstract: AMULET3 is the third fully asynchronous implementation of the ARM architecture designed at the University of Manchester. It implements the most recent version of the ARM architecture (v4T), including the Thumb instruction set. Significant architectural changes from its predecessors help achieve higher performance without sacrificing the advantages of asynchronous design. One of these changes is to incorporate a highly parallel instruction prefetch unit. This paper introduces the instruction prefetch unit in AMULET3, highlighting where speed-independent control circuits are implemented using signal transition graphs (STGs). In order to show how control circuits are implemented in the instruction prefetch unit of AMULET3, we present several examples with relevant STGs and the synthesized circuit results.

Keywords: asynchronous circuits, microprocessor, Petri nets, Signal Transition Graphs

1.   INTRODUCTION

AMULET3 is the third asynchronous implementation of the ARM architecture [1] to be produced at the University of Manchester. Its predecessors, AMULET1 [2] and AMULET2 [3], were intended to demonstrate that asynchronous circuits of this complexity are feasible and

171

A. Yakovlev et al.(eds.), Hardware Design and Petri Nets, 171-190.
© *2000 Kluwer Academic Publishers.*

practical; AMULET3 has been designed to be a commercially competitive macrocell. It is therefore required to deliver a performance similar to that of the contemporary synchronous ARM, the ARM9TDMI [4], and to implement the most recent version (v4T [5]) of the instruction set architecture including the 16-bit Thumb instruction set [6]. AMULET3 is being implemented in the same generic 0.35 μm 3 metal layer process as the ARM9TDMI. This implies a performance target of well over 100 MIPS (measured with Dhrystone 2.1), compared to the 40 MIPS delivered by AMULET2e on a 0.5 μm process.

Achieving this performance has necessitated a considerably different microarchitecture from the earlier AMULET processors. Most notable among the changes are the use of a Harvard architecture to increase memory bandwidth and the inclusion of a reorder buffer to handle data forwarding and memory faults. To cope with the former change, the instruction prefetch unit and the data interface are decoupled, whereas they were combined in the complex single address unit in AMULET2e. This paper is confined to describing asynchronous control circuit design in the instruction prefetch unit; readers having interests in other aspects of AMULET3 are referred to a related paper [11].

As in the previous AMULET processors, the architectural design is based on an asynchronous Micropipeline [7] structure using four-phase [8] control signals. All control circuits are developed on the basis of the speed-independent circuit assumption and this property is ensured using Petrify, an asynchronous synthesis tool [12]. Most of the control logic was specified with STGs [14][16], and each STG synthesized with Petrify. This paper does not show all of the control circuits implemented in the instruction prefetch unit but presents the general rules used to implement the control circuits including several examples.

In section 2, the specification of the instruction prefetch unit is given. This section is reproduced from a previous paper [11] with only minor changes and presents the context for the work reported here. The most recently updated information is added, such as the revised Figure 9.1. The top-level STG definition of the instruction prefetch unit is shown in section 3. Section 4 introduces general rules used in the control circuit design. In sections 5 to 7, several design examples are shown to explain how real control circuits are made. In section 8 simulation results are presented to show the performance of the instruction prefetch unit in terms of speed. The conclusions are given in section 9.

2.   INSTRUCTION PREFETCH UNIT

The instruction prefetch unit (Figure 9.1) is responsible for generating addresses for the instruction memory, which are sent via the Instruction Address Register ('IAR' in Figure 9.1) [13]. The instruction prefetch unit has a highly parallel organization, speculatively computing the outcome of all scenarios in parallel and then selecting the appropriate course of action in the final multiplexer. Although such speculation causes some unnecessary activity (and therefore wastes power) it is necessary here to meet the required throughput.

Usually the output addresses form an ascending sequence and are provided by a simple loop containing an incrementer ('INC'). When a branch occurs this loop may be interrupted asynchronously (via an arbiter) and loaded with a new address from the ALU. However, in AMULET3 several other functions are performed here also.

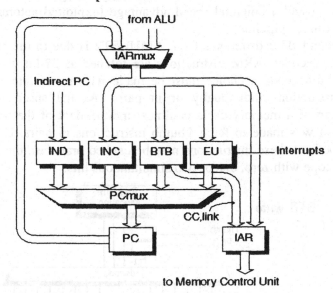

Figure 9.1. Instruction prefetch unit organization

2.1   Branch prediction

Branches disturb program flow and incur a considerable penalty in deeply pipelined systems. Sophisticated branch prediction mechanisms are now in use on state-of-the-art processors, but even a relatively simple branch predictor can significantly reduce pipeline disruption.

AMULET3 uses the same branch prediction mechanism as AMULET2 [9], namely a Branch Target Buffer ('BTB') which predicts a previously-taken invariant branch as 'always taken' until it is displaced from the BTB by a new entry. However there are two significant differences between the BTB in AMULET2 and that in AMULET3 [13].

The AMULET2 BTB records an address containing a branch instruction together with its target address. However, if a branch instruction address subsequently hits in the BTB the instruction is still fetched from memory and executed as it may be conditional and it may require a return address saving (if it is a BL – Branch-and-Link – instruction, used for procedure entry).

The information which AMULET2 gets from memory when it fetches a predicted branch amounts to only five bits (four condition bits and the 'L' bit). In AMULET3 these five bits are stored in the BTB so that the instruction does not have to be fetched in repeat encounters and the instruction memory may be bypassed. As branches account for 10%- 15% of ARM instructions [10], and the majority of these are cached in the BTB [9], this reduces the number of instruction fetches, yielding both a considerable power saving and a potential speed advantage (exploited automatically by the asynchronous pipeline).

The second BTB difference from AMULET2 is due to the presence of the Thumb decoder. ARM instructions are fetched as 32-bit words. When running Thumb code a choice must be made whether to fetch the 16-bit Thumb instructions individually or in pairs. As the speed and power consumption of a memory cycle is almost independent of the transfer size the decision was made to fetch Thumb instructions in pairs. However, as either or both of these instructions may be cached branches, the BTB must be able to cope with zero, one or two simultaneous hits.

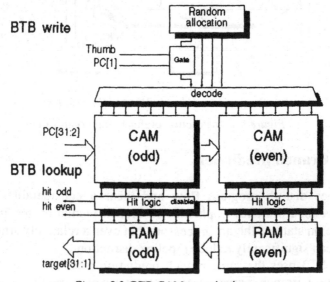

Figure 9.2. BTB CAM organization

This is achieved by splitting the BTB Content Addressable Memory (CAM) into two sections (see Figure 9.2). In Thumb mode each section works with one half word of the instruction pair; any potential conflicts are resolved by taking the 'even' half word (the Thumb instruction at the lower address) prediction because this will always be first in the instruction sequence.

When running ARM code the two sections are merged. This allows the BTB to cache a mixture of ARM and Thumb branches simultaneously without compromising the number of usable entries in either case.

2.2   Halting and interrupts

Most current CMOS technology dissipates very little power when not switching. This has been exploited in AMULET2e by causing the system to halt when no useful work can be performed, with demonstrable power-efficiency benefits [3].

Halting an asynchronous pipeline at any point soon causes the whole system to halt. Because there is no free running clock this reduces the number of transitions – and hence the power consumption – to near zero. In a synchronous system the clock oscillator could be stopped, but this is quite a complex procedure. The asynchronous system also recovers quickly (as there is no clock to restart). AMULET2e and AMULET3 exploit this by decoding a branch back to itself as a 'Halt' instruction and use this to stall the pipeline; this is fully compatible with much existing ARM code. The halt state is exited by the assertion of an enabled interrupt.

As alluded to above, the stall can occur anywhere within the pipeline. AMULET2, for example, stalls in the execution stage. AMULET3 adopts a somewhat cleaner model by stalling at the prefetch stage. This means that the processor restarts with an empty pipeline, which provides the fastest possible response, any 'rubbish' being cleared out at halt entry.

The interrupt signals are fed into the prefetch unit rather than the instruction decoder. This rather unusual feature provides both a clean interrupt model and a low interrupt latency. When an (enabled) interrupt is asserted it is arbitrated into the prefetch cycle and treated much like a predicted 'BL'. The interrupt 'hijacks' an instruction address, bypasses the memory, and proceeds down the execution path to save the return address. The PC is loaded with the address of the service routine (which is a constant, generated in the Exception Unit, 'EU' in Figure 9.1) in parallel and the prefetching of this code begins immediately.

A consequence of this approach is that the prefetch unit must store an up-to-date copy of the interrupt enable status. One danger is that this may be out of date because an operation already prefetched may change it. Another, related, problem is that the hijacked address may be in the 'shadow' of a branch and the interrupt may try to save an incorrect return address.

Both these problems are solved by treating control instructions (such as enabling/disabling interrupts) as branches, and branches as potential control instructions. This is not particularly onerous because almost all instructions which can alter these flags (e.g. software interrupts, return from interrupts, etc.) also cause flow changes anyway. If an interrupt has occurred in a branch shadow it will be discarded in the same way as any other erroneously prefetched instruction. Concurrently, the branch will reach the prefetch unit, re-enable interrupts, and immediately cause the interrupt entry mechanism to repeat, this time saving the branch target as the return address.

2.3   Indirect branches

ARM programs often load the Program Counter (PC) directly from memory as part of a subroutine return (and, less frequently, as a result of a jump table lookup). Typically a subroutine return restores the PC together with a set of working registers using a load multiple (LDM) instruction. The load ordering is such that the lowest numbered register is loaded first, and thus the PC (R15) is loaded last. This delays the start of instruction fetching from the return address and compromises performance.

AMULET3 incorporates an optimization which exploits the separate instruction memory port. The execution unit passes the load address of the PC value back to the prefetch unit via the branch address path in parallel with initiating the other register transfers in the data interface. This 'branch' terminates prefetching from the redundant instruction stream and prompts a single read cycle which fetches the new PC. This is then returned to the prefetch unit (via 'IND' in Figure 9.1) where instruction fetching resumes. With a typical subroutine return much of this should happen whilst the data transfers are proceeding and so the new instructions should be available before the LDM has completed.

Note that this feature imposes a significant constraint on the memory designer: the instruction and data memories must be coherent because a PC value is stored via the data port and then read via the instruction port. The first AMULET3-based system has a unified memory which is dual-ported to give independent instruction and data ports. Coherence is therefore not an issue here.

3.   CONTROL PATH OVERVIEW

The control path in the instruction prefetch unit can be viewed as a large arbitration block including a fork and join connection as shown in Figure 9.3. Note that Figure 9.3 is a very simplified STG of the instruction prefetch unit from a high-level viewpoint. Furthermore, this STG uses a more abstract labelled Petri net when compared to the original STG definition. We labelled each transition according to a logical function in the instruction prefetch unit, not a detailed control signal transition. However, other STGs used in this paper are based on the original STG definition.

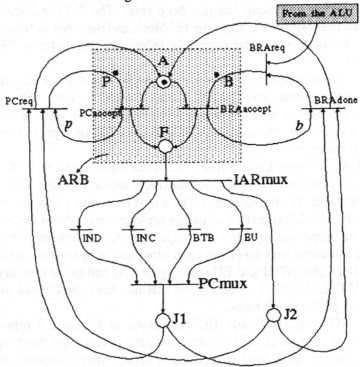

Figure 9.3. Top-level STG of the instruction prefetch unit

Arbitration occurs between two requests; one is PCreq and the other is BRAreq (see Figure 9.3). The fork is situated in the IARmux and the join in the PCmux (see Figure 9.1 and Figure 9.3).

Each dot in Figure 9.3 represents an initial token when the circuit is in the reset state and each bar represents the behaviour of each block in Figure 9.1.

The name of each transition is the same as the corresponding block name in Figure 9.1 except the PCreq, BRAreq, PCaccept, BRAaccept and

BRAdone transitions. The PCreq transition represents the recirculating request from the incremented PC value which is generated entirely within the instruction prefetch unit. The BRAreq transition represents a request for a new value (a branch target PC) from outside the instruction prefetch unit. This new value will replace the old value in the internal recirculating loop.

The box labelled 'ARB' represents an asynchronous arbiter. The token in place A enables either the PCaccept or the BRAaccept transition to fire (assuming their respective requests are active), but only one at a time. When both requests fire at nearly the same time the arbiter ensures that one, and only one, is granted.

Assume that the circuit has just been reset. The STG in Figure 9.3 has two initial requests; one is from the PC block and the other is from the ALU block (see Figure 9.1). The former is PCreq and the latter is BRAreq in Figure 9.3.

When the arbitration takes place (in ARB) either the PCaccept or the BRAaccept transition may fire. No matter which one fires, the other can't fire until the token is back in place A. That is, arbitration starts when place A sends a token to one of transitions; PCaccept or BRAaccept, and ends when a token is put in place F.

Next, place F gets a token from one of the input transitions, PCaccept or BRAaccept. The token enables the transition named IARmux in Figure 9.3, which represents the instruction prefetch unit receiving a new address either from the PC block inside the instruction prefetch unit or from the ALU block outside the instruction prefetch unit, depending on the arbitration result. This IARmux transition fires tokens to each block in the instruction prefetch unit such as IND, INC, BTB and EU (see Figure 9.1) and to the memory control unit in AMULET3. At the moment the IARmux fires, the control path in the instruction prefetch unit forks.

The IND, INC, BTB and EU transitions in Figure 9.3 represent the behaviour of each block in the instruction prefetch unit and these transitions are described in the following sections in detail. After these transitions fire, four tokens are joined at the transition named PCmux. The behaviour of the PCmux is to select one of the addresses from IND, INC, BTB or EU for the next program counter and then store this address in the program counter address flip/flop.

The places named J1 and J2 play the role of feeding tokens to either the PCreq transition or the BRAdone transition. Whether tokens go to PCreq or BRAdone is decided by the arbitration that happened in the ARB. If BRAaccept fires, the arrow b feeds a token to BRAdone and if PCaccept fires, the arrow p provides a token to PCreq. When either PCreq or BRAdone fires, a token is put back in place A. This means that all of the arbitration behaviour is finished in the instruction prefetch unit and a new arbitration can occur between PCreq and BRAreq. Note that the BRAreq transition can fire only when the ALU sends a new branch PC to the instruction prefetch unit.

4.   FOUR-PHASE CONTROL DESIGN STRATEGY

Since the 4-phase broad protocol is used for the AMULET3 design, there is a return-to- zero phase to recover the control signal as shown in Figure 9.4. This phase could be a disadvantage in terms of speed since the phase does nothing but the return-to-zero action. When a designer tries to implement dynamic datapath circuits, however, this return-to-zero phase could be usefully employed for precharging the circuits.

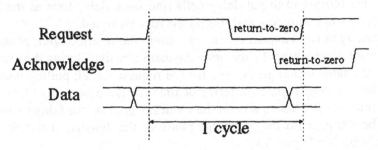

Figure 9.4. 4-phase broad protocol

If we view the request signal (rising edge) in Figure 9.4 as the arrow from place F in Figure 9.3, and the acknowledge signal (falling edge) in Figure 9.4 as the arrow to place A either from PCreq in Figure 9.3 or from BRAdone in Figure 9.3, one cycle time in the instruction prefetch unit is defined from the rising edge of the request signal in Figure 9.4 to the next rising edge of the same signal. The AMULET3 specification requires this cycle time to be less than 7 nanoseconds and this guarantees that the final chip speed will be well over 100 MIPS which is comparable with the latest synchronous ARM microprocessors.

The strategy used to design control circuits depends on the manner in which the datapath circuits are implemented; static or dynamic. When static datapath circuits are used, the control circuit design is rather straightforward since the AMULET3 design adopts the bundled data scheme, where completion signals are generated by mimicking the worst-case time consumed in the datapath. That is, delay elements are inserted into the path of the request signal and the resulting delayed request signal provides the request signal to the next block as shown in Figure 9.5.

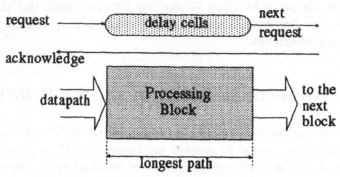

Figure 9.5. Control scheme for a datapath

What is needed is to detect the longest path (in terms of time) of the processing block and to put delay cells (the same delay time as the longest path plus a silicon process margin) on the request signal.

When dynamic datapath circuits are used, the return-to-zero phase could be used for precharging the dynamic datapath circuits. During the evaluation phase, the same technique is used for the request signal; putting delay cells to mimic the longest path (in terms of time) of the processing block. When the return-to-zero phase starts on the request signal (at the falling edge), this could be a trigger for the precharge phase of the dynamic datapath circuits (see scheme 2 in Figure 9.6).

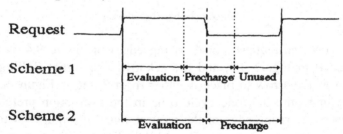

Figure 9.6. Dynamic circuit control timing

The EU block in Figure 9.1 is comprised only of static datapath circuits and therefore the return-to-zero phase of the request signal performs no useful work. The BTB and INC blocks consist of dynamic datapath circuits. For these blocks the return-to-zero phase could be used to perform the precharge phase.

However, even though dynamic datapath circuits are used, the return-to-zero phase is of no use if a fully decoupled control circuit [8] is used to give a highly concurrent circuit operation. In general higher concurrency might be expected to make a faster circuit, since as soon as the request signal is transferred to the processing block, the acknowledge signal of the block is issued back to the previous block (without waiting for the acknowledge signal from the next block) and the next request signal is transferred to the next block concurrently.

Does higher concurrency always mean a faster control circuit for dynamic datapath circuits? The answer is no in our design.

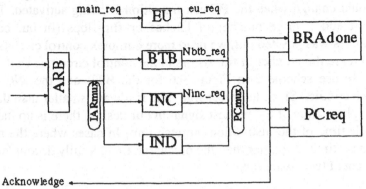

Figure 9.7. Request and acknowledge signals in the instruction prefetch unit

If we use fully decoupled control circuits for the INC and BTB blocks, we can use control timing following scheme 1 in Figure 9.6. This looks faster than scheme 2 in Figure 9.6. However, there are two disadvantages in terms of the speed and size of circuits.

Let's consider the speed issue first. As shown in Figure 9.7, the INC and BTB transitions occur in the middle of the instruction prefetch unit between the IARmux and the PCmux transitions and the cycle of the instruction prefetch unit is finished when either the PCreq transition or the BRAdone transition sends an acknowledge signal (falling edge) back to the arbiter. The BRAdone and PCreq transitions are implemented at the IAR and PC blocks in Figure 9.1 respectively.

The cycle time in the instruction prefetch unit depends on how fast the request signal can reach the BRAdone and PCreq transitions. If we use control scheme 1 in Figure 9.6 for the INC and BTB blocks, a more complex circuit is needed to implement a fully decoupled circuit, since the fully decoupled circuit must retain some control state. The state is needed for two possible cases. One is when the precharge phase is finished before a new request rising edge arrives at the control circuit and the other is the reverse case. In the former case, the information that the precharge phase is finished must be memorized until the new request rising edge enters the control circuit and vice versa in the latter case. The more complex circuit will reduce the speed of forwarding the request signal from the main_req to the BRAdone and PCreq transitions in Figure 9.7. Since the INC and BTB blocks are in the middle of the instruction prefetch unit, there is no advantage if the rest of blocks are not finished, even though these blocks finish their functions.

Secondly, consider the size issue. The PCmux transition will happen after the EU, BTB, INC and IND transitions are finished as shown in Figure 9.3. In order to use a fully decoupled control circuit, we need storage elements in the datapath circuits at the ends of the BTB and the INC blocks. Otherwise data could change when the PCmux transition is being activated. Therefore we must add a large number of latches or flip-flops (in our case 31x2 latches). As was pointed out, we need more complex control circuits and this will increase the number of transistors in the control circuit as well.

So we use scheme 2 in Figure 9.6 for the BTB and the INC dynamic datapath circuits. Since the precharge time is always smaller than the return-to-zero phase time of the request signal in our design, there is no increase on the cycle time of the instruction prefetch unit. In cases where the return-to-zero phase time is smaller than the precharge time, a fully decoupled control circuit could be considered.

5.   THE DESIGN OF THE INC BLOCK

The behaviour of the INC block is shown in Figure 9.8, where the main_req and the Ninc_req signals are as same as in Figure 9.7. The Nprech signal controls the precharge phase of the INC block and the Ncomp signal indicates the evaluation phase of the INC block is finished (completion signal).

Since we choose to use scheme 2 in Figure 9.6, we have a serial forward connection from the main_req to the Ninc_req as shown in the first line of Figure 9.9. Note that Nprech, Ncomp and Ninc_req are active low signals and main_req is an active high signal. However, we can use a decoupling technique for the precharge phase as shown in the second and third lines of Figure 9.9. The second line shows that the precharge phase starts immediately after main_req- (triggering the precharge phase of the INC datapath circuit) enters the INC block and the third line shows that the return-to-zero phase of the request signal is transferred to the next block (PCmux in Figure 9.7) at the same moment as the precharge phase starts. (This is called semi-decoupling [8].) Therefore we can reduce the time needed for the return-to- zero phase of the request signal.

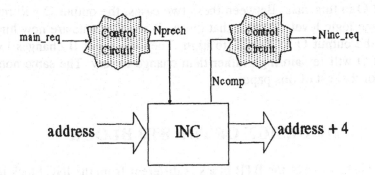

Figure 9.8. INC control circuit

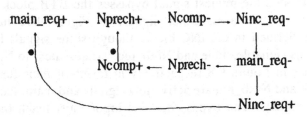

Figure 9.9. STG for the INC block

The synthesized control circuit of the STG in Figure 9.9 is shown in Figure 9.10, where a C- gate is used.

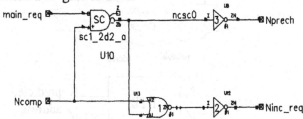

Figure 9.10. Schematic of the INC control circuit

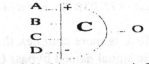

Figure 9.11. 4-input asymmetric C-gate

The C-gate is widely used in asynchronous design. If we have a C-gate as in Figure 9.11, the behaviour is as follows. When the A, B and C inputs are high, the output O will change to high. In this case the input D does not affect the change of the output O. When the B, C and D inputs are low, the output O will change to low. The input A does not have an impact on the output O in this case. Between these two cases, the output O will remain at the same logic level. Assume that the A, B and C inputs are now high. This leads the output O to change to high. Then the input B changes low. The output O will remain high rather than change to low. The same notation is used for the rest of this paper.

6.   THE DESIGN OF THE BTB BLOCK

The behaviour of the BTB block is different from the INC block in that it has a Boolean input signal labelled bypass (see Figure 9.12). This bypass signal can be used when the BTB block is not to be turned on. When this signal is activated the request signal bypasses the BTB block and the BTB operation is not activated. Other signals in Figure 9.12 have the same functions as defined in the INC block, though some signals have different names such as Nenable, done and Nbtb_req   (equivalent to Nprech, Ncomp and Ninc_req in Figure 9.8 respectively in terms of their functions). Note that Nenable and Nbtb_req are active low signals and main_req and done are active high signals. The bypass boolean input must reach the BTB block before main_req and must maintain its logic level until the return-to-zero phase of the Nbtb_req is finished.

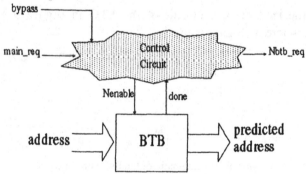

Figure 9.12. BTB control circuit

The STG of the BTB block is more complex than that of the INC block since there is a boolean logic signal named bypass (see Figure 9.13). The P0 and P1 places and the bypass+ and bypass- transitions are to model the boolean input signal. Depending on the logic level of the bypass signal, when main_req enters the BTB block the token in place P3 will go either to the Nbtb_req- transition or to the Nenable- transition. The former is the case when bypassing the BTB block and is a straightforward 4-phase protocol. The latter is the same behaviour as used in the INC block as shown in Figure 9.9.

The synthesized result for the BTB control circuit is shown in Figure 9.14.

In Figure 9.14, the done signal is forked; one fork going to a complex gate and the other to an inverter. The inverted done signal is then used to drive an input to a C-gate. This may violate the conditions of speed-independence [15]. However, the control circuit is synthesized locally (ensuring that the logic components of the circuit are laid out closely) and simulation of the layout confirms that the circuit works correctly. The same procedure is used to check the speed-independence assumption throughout the control circuit design whenever there is an inverted input.

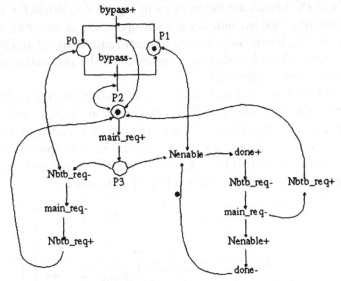

Figure 9.13. STG for the BTB control circuit

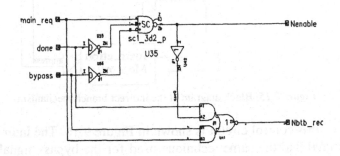

Figure 9.14. Schematic of the BTB control circuit

7.   THE DESIGN OF THE EU AND IND BLOCKS

The EU block has static datapath circuits and the control circuits are directly implemented using the delay matching method explained in section 4.

The IND transition in Figure 9.3 is chopped off and the function is merged into place J1 in Figure 9.3 (equivalent to the PC block in Figure 9.1), since indirect program address loading is not always initiated but when the boolean signal 'indir' at the control circuit in Figure 9.15 is activated. The indirect branch mechanism is shown in the block diagram of Figure 9.15. The IAR and PC blocks are the same as in Figure 9.1. When the indir signal is high (meaning that an indirect branch operation is needed), the control circuit for the PC block waits for the indirect request signal and the bundled indirect channel. This bundled data comes from the instruction memory by indirect addressing via the IAR.

After finishing the indirect channel communication, the main_ack goes back to the ARB in Figure 9.3. This behaviour can be explained as follows. The token in place J1 of Figure 9.3 goes to either the PCreq transition or the BRAdone transition depending on the result of the arbitration in ARB. However, when an indirect branch is required, the PCreq and the BRAdone transitions can't fire until the indirect branch is finished.

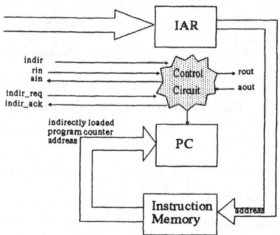

Figure 9.15. Block diagram of the indirect branch mechanism

The STG of this control circuit is shown in Figure 9.16. The indir signal is a boolean signal and the same technique used for the bypass signal in Figure 9.13 is exploited again in Figure 9.16. Therefore the change of the indir signal must arrive at the control circuit before the request signal 'rin' reaches the circuit and must be maintained until the return-to-zero phase of the acknowledge signal is finished. The semi-decoupled technique is adopted for this control circuit to boost speed when rout+ in Figure 9.16 is activated. At the same moment that rout+ is invoked to send a request signal to the next block, ain+ happens to return an acknowledge back to the previous block. There is no need for ain+ to wait for the aout+ response from the next block.

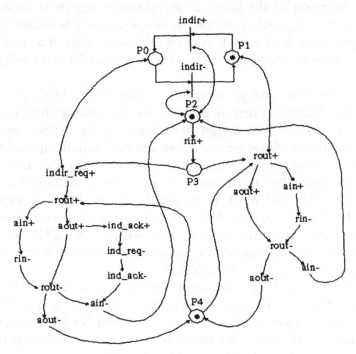

Figure 9.16. STG for the PC block

The synthesized result for the PC control circuit is shown in Figure 9.17.

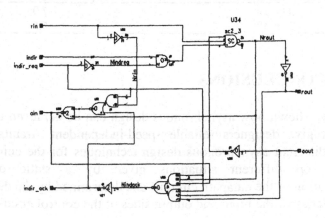

Figure 9.17. Schematic of the PC block

8.   SIMULATION RESULTS

Now, we present the performance figures of the instruction prefetch unit in terms of speed.

The definition of the speed in asynchronous design is different from synchronous design since synchronous design can only be measured in terms of the worst case. In synchronous design the worst delay of a block in a chip defines a fixed global clock cycle time regardless of different delays in each block.

In asynchronous design, however, the same logic block can finish its process in different times depending on its running function. In our instruction prefetch unit design two cases can be defined by different running conditions. Since we use a fork and join form in our design, forked blocks must finish their processes before the join. The EU, INC and BTB blocks must finish their processes before the PCmux transition is activated in Figure 9.3. (The IND block is merged in the PC block as explained in section 7.) The EU and the INC blocks must be used every time for the function of the instruction prefetch unit, whereas the BTB block can be turned off by a user. Therefore we can have two operation modes in our performance simulation as shown in table 1.

The normal case in the instruction prefetch unit is that the BTB block is turned on. When an indirect branch happens the cycle time will be longer than the normal case since more time is required for the indirect channel communication. However, this condition is not included in the performance test since it depends on the speed of the instruction memory.

Table 1. Cycle time in the instruction prefetch unit

Operation modes	Cycle time (nanoseconds)
BTB on	7.0
BTB off	5.5

9.   CONCLUSIONS

We have shown how asynchronous design can be achieved using STGs, which can give designers reliable speed-independent circuits. We have exploited different asynchronous design techniques for the control circuits depending on different situations given by a static or dynamic implementation of the datapath circuits, and determined by the decoupling of the dependency of the input and output sides of the control circuit.

10.   ACKNOWLEDGMENTS

The development of AMULET3 was supported primarily within the EU-funded OMI- DE2 and OMI-ATOM projects, and the authors are grateful to the European Commission for their continuing support for this work. ARM Limited coordinated these projects; their support, and that of the other project partners, is also acknowledged.

The VLSI design work has leant heavily on CAD tools from Compass Design Automation (now part of Avant!) and EPIC Design Technology, Inc. (now part of Synopsys).

11.   REFERENCES

[1]   Furber, S.B., "ARM System Architecture", Addison Wesley Longman, 1996. ISBN 0-201-40352-8.

[2]   Woods, J.V., Day, P., Furber, S.B., Garside, J.D., Paver, N.C., Temple, S., "AMULET: An asynchronous ARM microprocessor", IEEE Transactions on Computers, Vo. 46, No. 4, pp. 385-398, April 1997.

[3]   Furber, S.B., Garside, J.D., Riocreux, P., Temple, S., Day, P., Liu. J., Paver, N.C., "AMULET2e: An Asynchronous Embedded Controller", Proceedings of the IEEE, volume 87, number 2, pp. 243-256, February 1999.

[4]   Segars, S., "The ARM9 Family - High Performance Microprocessors for Embedded Applications", Proc. ICCD'98, Austin, October 1998, pp. 230-235.

[5]   Jaggar, D., "Advanced RISC Machines Architecture Reference Manual", Prentice Hall, 1996. ISBN 0-13-736299-4.

[6]   Segars, S., Clarke and Goudge, "Embedded Control Problems, Thumb, and the ARM7TDMI", IEEE Micro, **15** (5), October 1995, pp. 22-30.

[7]   Sutherland, I.E., "Micropipelines", Communications of the ACM, **32** (6), June 1989, pp 720-738.

[8]   Furber, S.B. and Day, P., "Four-Phase Micropipeline Latch Control Circuits", IEEE Trans. on VLSI, 4 (2), June 1996, pp. 247-253.

[9]   York, R., "Branch Prediction Strategies for Low Power Microprocessor Design", M.Sc. Thesis, University of Manchester 1994.

[10] Jagger, D., "A Performance Study of the Acorn RISC Machine", M.Sc. Thesis, University of Canterbury, 1990.

[11] Garside, J.D., Furber, S.B., Chung, S.-H., "AMULET3 revealed", Accepted for the 5th International Symposium on Advanced Research in Asynchronous Circuits and Systems, Barcelona, 19-21 April 1999.

[12] J. Cortadella et al, "Petrify: a Tool for Manipulating Concurrent Specifications and Synthesis of Asynchronous Controllers", IEICE Transactions on Information and Systems, E80-D(3): 315-325, 1997.

[13] Chung, S. -H., "The Design of the Branch Target Cache for an Asynchronous Microprocessor", M.Phil. Thesis, University of Manchester, October 1998.

[14] Chu, T-.A., "On the Models for Designing VLSI Asynchronous Digital Circuits", Integration, the VLSI journal, 4(2): 99-113, June 1986.

[15] K. van Berkel, "Beware the Isochronic Fork", Integration, the VLSI journal, vol. 13, pp. 103-128, 1992.

[16] Rosenblum, L. Y., Yakovlev, A. V., "Signal graphs: from self-timed to timed ones", Proc. of the Int. Workshop on Timed Petri Nets, Torino, Italy, July 1985, IEEE Computer Society Press, NY, 1985, pp. 199-207.

IV

HARDWARE DESIGN
METHODS AND TOOLS

Chapter 10

ELECTRONIC SYSTEM DESIGN AUTOMATION USING HIGH LEVEL PETRI NETS

Patrik Rokyta
Siemens AG Munich,
Germany
Patrik.Rokyta@icn.siemens.de

Wolfgang Fengler and Thorsten Hummel*
Ilmenau Technical University,
Germany
{Wolfgang.Fengler,Thorsten.Hummel}@theoinf.tu-ilmenau.de

Abstract   A design and implementation methodology for system specification, modelling and implementation using a special kind of high level Petri nets is described. Electronic system design automation tools are used to generate synthesizable VHDL code from a Petri net model. For the design of large systems with regular structures the use of coloured Petri nets will improve the handling and flexibility. Two design examples illustrate the described methodology.

Keywords: Hardware Design, Petri nets, VHDL

1.   INTRODUCTION

A design and implementation methodology must provide tools for system specification, modelling and implementation.

---

*Supported by DFG grant GRK 164/1-96

A. Yakovlev et al.(eds.), Hardware Design and Petri Nets, 193-204.
© *2000 Kluwer Academic Publishers.*

The expected behaviour of the system (behavioural model) is captured in the specification. This model can be used to check the algorithm of the system without any constraints towards implementation.

A formal language is used to capture a RTL model of the design. The most popular languages for capturing a RTL model are Verilog and VHDL [4]. These hardware description languages can be synthesized to a gatelevel netlist.

Petri nets have shown to be a powerful formal language to specify and model the behaviour of parallel systems at a high abstraction level. Since a PN can be analysed, the error detection is possible without any verification methods.

Electronic System Design Automation (ESDA) allows the designer to capture a graphical model of the design using state diagrams, flow charts or truth tables. The creation of a synthesizable VHDL model is provided by a built-in VHDL generator.

Since Petri nets are more flexible for modelling parallel designs they should be used instead of state machines. Using Petri nets the merger between the behaviour and the state machine can be fulfilled. The analysis of the design can be extended to its behaviour (i.e. reachable values of the output signals instead of reachable state vectors).

The VHDL generator must extract a fully synthesizable VHDL code out of any type of Petri net model.

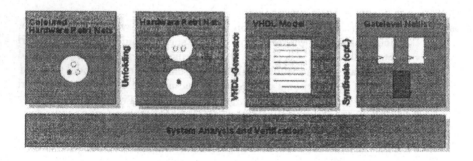

Figure 10.1   ESDA based System Design using Hardware Petri Nets.

2.   THE HARDWARE PETRI NET

In order to use Petri nets for capturing a design for verification or a design for manufacturing, the Place Transition Net must be extended for purposes which allow high level signal modelling. This kind of Petri net is refered to as the Hardware Petri Net (HPN):

- The Places store a certain number of tokens up to their capacity. This feature allows modelling of bus signals. The capacity of the place must correspond with the width of the bus.

- When a place stores at least one token the corresponding asynchronous I/O-function, which is attached to the place, is executed. This is supposed to be used for modelling of asynchronous behaviour. The function is defined using VHDL statements and must be fully synthesizable.

- When a transition fires, the corresponding synchronous I/O-function, which is attached to this transition, is executed. This is useful for modelling of synchronous behaviour. The function is defined using VHDL statements and must be fully synthesizable.

- Besides known arcs, enabling and setting arcs are defined.
  Enabling arcs define an additional firing condition related to the marking of the place the arc is connected with. Any condition is allowed on this arc (e.g. < 3, $>= 4$). An inhibitor arc, which is included in the enabling arcs, provides an enabling function, when the place stores no token ($= 0$).
  Setting arcs cause the place to obtain a certain number of tokens, which is defined on this arc.
  Since a place is represented by a flip-flop in a logical design, the setting arc can be used to synchronize external signals with the active edge of the system clock.

- An additional external firing condition can be attached to any transition. Though, this can cause timing violations at the flip-flops.

Using HPN for signal modelling allows the modelling of either synchronous or asynchronous behaviour. Setting the corresponding default value for asynchronous signals, a latch, a tri-state or a combinational logic design can be synthesized. A reset value can be defined for any signal of the design which takes place if reset is active.

Using I/O-functions either on places or on transitions, a Mealy machine is synthesized. For a critical design using places with the corresponding capacity (1 for 1 bit, 3 for 2 bits, 15 for 4 bits) to represent the value of the output signals, a Moore machine is synthesized.

Firing transitions are synchronized with the active edge of a clock.

3.   HIERARCHICAL DESIGN AND PARTITIONING

The state vector of a HPN is built out of bit vectors where each of them corresponds with the capacity of the related place. Since the design can have an unlimited number of places with unlimited (the highest integer used in the synthesis tool) capacity, the state vector can become very large. Since the FSM optimization does not work with large state vectors, the design must be split into several (hierarchical) blocks. The FSM optimization (reachable states, encoding style) can be applied to the lowest level of the hierarchy.

The partitioning of a HPN design is fulfilled at the top level. In this case the design is split into several units. Each unit is synchronized with individual clock and reset.

Within each unit a hierarchical HPN can be captured using macros. A macro allows to descent to the next hierarchy level. Since the hierarchy is directly translated to a VHDL model, incremental synthesis is possible.

For each signal within a unit only one driver is synthesized. To get more drivers for one signal, the signal must be modeled in more than one unit and must be defined as a tri-state signal.

4.   GENERATION OF SYNTHESIZABLE VHDL-CODE

Using HPN a generation of synthesizable VHDL-Code is possible for any kind of design. Using the reachable states the synthesis result can be optimized since the lowest level of the hierarchy is always a FSM.

The top level of a HPN design consists of the unit instances. Since a unit requires a synchronous reset, the reset is synchronized outside this unit with the unit clock and is handled as an asynchronous reset within this unit in order to save logic and get more efficient synthesis results.

Each VHDL model of a HPN unit consists of the synchronous and the asynchronous I/O-process where all I/O-functions, which are attached to the places and the transitions of this unit, are executed. The execution order is controlled by the priority value attached to these elements.

The most important part within a HPN unit is the FSM instance which simulates the firing of transitions. This FSM is a separated VHDL model. In this model the firing conditions of the transitions are calculated. A special VHDL process detects and removes any firing hazards. Another VHDL process controls the token flow within the unit. Since the FSM can be a hierarchical design using macros it can consist of other instances of FSM at the lower level.

```
firing conditions:

t1 <- (p1 <- 15);
t2 <- (p1 >- 3);
t3 <- (15 <- 20); -- TRUE
t4 <- (p1 > 4);
```

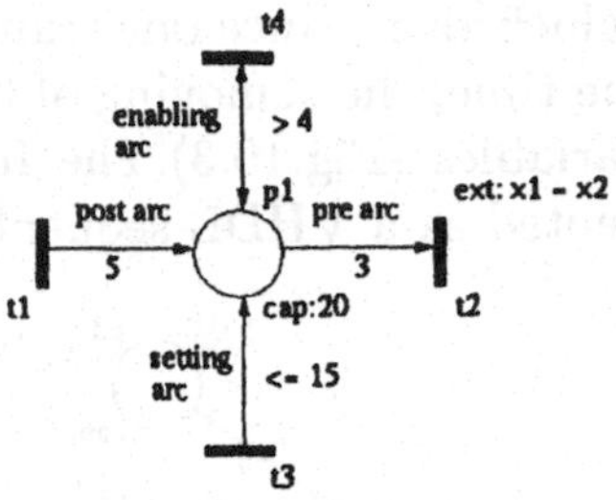

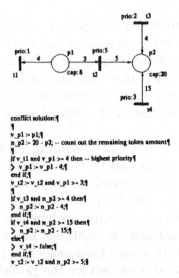

Figure 10.2 Firing condition of the transition.

The firing conditions of the transitions are calculated out of the markings of the connected places and the token value of the related arcs. The firing condition is extended by the external firing condition which can be attached to the transition. Enabling arcs and setting arcs affect the firing condition of the transition as well (Fig.10.2). The token value of the setting arc must not exceed the capacity of the related place element.

```
conflict solution:¶
¶
v_p1 :- p1;¶
n_p2 :- 20 - p2; -- count out the remaining token amount¶
¶
if v_t1 and v_p1 >- 4 then -- highest priority¶
> v_p1 :- v_p1 - 4;¶
end if;¶
v_t2 :- v_t2 and v_p1 >- 3;¶
¶
if v_t3 and n_p2 >- 4 then¶
> n_p2 :- n_p2 - 4;¶
end if;¶
if v_t4 and n_p2 >- 15 then¶
> n_p2 :- n_p2 - 15;¶
else¶
> v_t4 :- false;¶
end if;¶
v_t2 :- v_t2 and n_p2 >- 5;§
```

Figure 10.3 Removing of firing hazards.

Firing hazards are removed by defining a firing priority of the transitions which are conflicting. Only transitions with the highest priority will fire at the specified clockedge. Since one transition can take part in more firing hazards at one time, the removing of the firing hazards must be done using VHDL variables (Fig.10.3).The functionality of solving firing hazards is implemented as a VHDL sequential process.

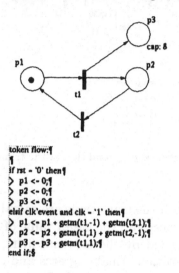

```
token flow:¶
¶
if rst = '0' then¶
 > p1 <- 0;¶
 > p2 <- 0;¶
 > p3 <- 0;¶
elsif clk'event and clk = '1' then¶
 > p1 <- p1 + getm(t1,-1) + getm(t2,1);¶
 > p2 <- p2 + getm(t1,1) + getm(t2,-1);¶
 > p3 <- p3 + getm(t1,1);¶
end if;§
```

Figure 10.4   Token flow with active clock edge.

The token flow is controlled by another VHDL process. It consists of the detection of the reset phase and the active clock edge. During reset the marking of the place elements is set to their start value. If the active clock edge is detected the token flow will takes place (Fig.10.4). The user defined function 'getm' returns the token amount which will relatively change the marking of the related place if the corresponding transition fires.

Using I/O functions the optimal scheduling must be done in advance. Since scheduling is not supported by Petri nets, it must be done manually by synthesizing examples which correspond to the functionality of the design. The synthesis results in the number of multicycles or the scheduling plan for the corresponding arithmetic operation. Scheduling can be done using formal analysis methods as well - synthesizing test examples does relate to the target technology and includes the timing of the wiring and the total area needed for the gatelevel design.

Since adding or substracting of tokens is an arithmetic operation as well, the clock period must be set to a value, where the marking of the places can be calculated at once. Using multicycles is not possible at this point. For time critical designs, the design can be turned to a finite state machine and FSM optimization can be run. This will remove all arithmetic operations from the token flow process. The design bases on changes, which are done to the state vector of the design. The state vector is built out of the marking of all places in the design. FSM optimization is supported by the synthesis tool. The reachable markings should be known to allow an extended optimization of the state vector and to save the amount of the state registers.

Allocation, structuring, flattening and mapping to the target technology is done by the synthesis tool i.e. DESIGN_ANALYZER (Synopsys, Inc.) or AUTOLOGIC (Mentor Graphics, Inc.) since the VHDL code generated out of a Hardware Petri Net uses full synthesizable RTL (Register Transfer Level) VHDL subset. Behavioural synthesis is not necessary.

5. UNFOLDING COLOURED PETRI NET DESIGNS TO HPN

Using ordinary Petri nets for modeling of large designs the resulting nets often seem to be badly arranged and confused. But typically, they frequently contain regular subnet structures. Using coloured Petri nets [6] simplifies the net structure by enfolding, i.e. information of the net structure will be transferred into the description of the net elements. The entire modeling ability will be kept but the handling and flexibility will be improved.

For simulating and generating the VHDL code the coloured HPN has to be unfolded, i.e. the information contained in the description of the net elements will be transfered back into a net structure without any loss of information. There are existing tools [10] including unfoulding rules to automate these processes.

6. DESIGN EXAMPLES

6.1 A MOTOR SUPERVISOR

This device controls a motor [3]. The motor is moved to a specific angle which is defined by the user. The motor will move in that direction which is shorter to reach the specific angle. The motor receives pulses, where each pulse means a movement of 1 degree. The new angle is defined by PHI and will be set using the signal LOAD (Fig.10.5). The supervisor stores the old angle in the variable OLDPHI and counts out the difference (DIFF) and the direction DIR. Once these values are counted out, the supervisor sends the corresponding pulses toward the motor.

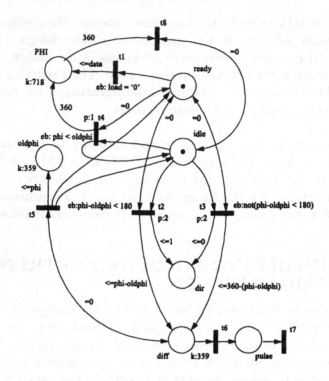

Figure 10.5 Motor Supervisor - Petri net model.

The functional simulation shows motions from 0 to 5, from 5 to 350, from 350 to 349, and from 349 to 0 degrees (Fig.10.6).

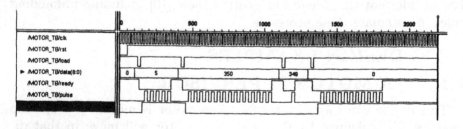

Figure 10.6 Motor Supervisor - Functional Simulation.

The synthesis results are constrained by area since the timing is not the point of interest.

Using the lca300kv target library (LSI Logic), following results are achieved:

15 ports, 393 nets and 293 cells.

The synthesized area amounts to:
1813 (total), 1304 (cells) and 509 (nets).

6.2 A SIMPLE ATM CELL RATE POLICER

Using STM (synchronous transfer mode) a channel with fixed transmission rate is reserved for each connection. ATM (asynchronous transfer mode) allows connections with variable transfer rate. This rate can be defined by the user and should not be crossed at any time. To detect any violations a policer is necessary to watch over the actual rate of an ATM connection. If a violation is detected, extra fee has to be paid by the user or the connection will be refused. Because some devices can cause a higher cell rate than allowed (multiplexers) a temporary crossing of the specific rate should be allowed.

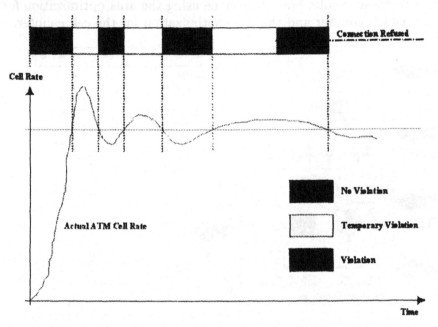

Figure 10.7 Simple ATM cell policer - Algorithm.

The ATM cell rate policer uses the leaky bucket algorithm [2]. It consists of two parallel working parts - the decrement counter (ZT) and the cell counter (Z).

Each time an ATM cell is received (ATM) the cell counter increments. If the number of cells exceed the allowed limit (S) the acknowledge of the incoming cell will be denied (ACK). The decrement counter decrements the cell counter each T period. The decrement value is specified by D.

The period and the decrement value specifies the time where crossing the reserved cell rate is allowed.

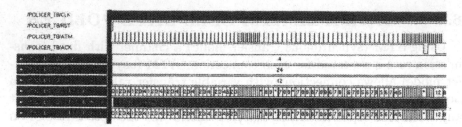

Figure 10.8   Simple ATM cell policer - Simulation.

The synthesis results have been done using the area optimization for the decrement counter and the time optimization for the cell counter.

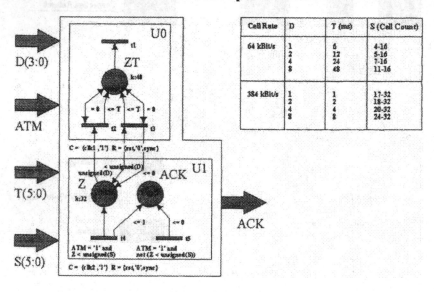

Cell Rate	D	T (ms)	S (Cell Count)
64 kBit/s	1	6	4-16
	2	12	5-16
	4	24	7-16
	8	48	11-16
384 kBit/s	1	1	17-32
	2	2	18-32
	4	4	20-32
	8	8	24-32

Figure 10.9   Simple ATM cell policer - HPN model.

The ATM cell rate policer is necessary for each ATM channel. Using a coloured Petri net the entire functionality can be flexibly captured in the same Petri net as shown in this example.

7. SUMMARY AND CONCLUSIONS

Using Petri nets to specify and model a digital system is an improvement in comparison to other ESDA capturing methods e.g. state machines. Using the analysing methods and the capability to describe parallel tasks, a powerful method is found to detect a large number of design errors prior to system implementation.

Since the VHDL generation out of a Hardware Petri Net is possible, a synthesizable design (e.g. FPGA or ASIC) can be captured at a higher abstraction level than VHDL. The creation of an executable specification even for large systems is possible as well.

To enlarge the flexibility of Petri nets in a hardware design process, the system should be captured as a coloured HPN and unfolded to a HPN before generating the VHDL code. Both steps unfolding a CHPN and the VHDL code generation can be done automatically.

References

[1] Carlson, S.: *Introduction to HDL-Based Design using VHDL*. Synopsys Inc., 1991.

[2] G. Daisenberger, J. Oehlerich, G. Wegmann: Two concepts for overload regulation in SPC switching systems: STATOR and TAIL. *Telecommunication Journal*, vol.56, V/1989.

[3] W. Fengler, A. Karg: Design of complex embedded systems based on different Petri-net interpretations. In *Advanced Simulation Technologies Conference*, Boston, MA, 5.-9.4.1998.

[4] *IEEE Standard VHDL Language Reference Manual - IEEE Std 1076-1993*. The Institute of Electrical and Electronics Engineers Inc., 1994.

[5] *The System Specs VHDL Generator - Beta2-Version*. Data Sheet. Ivy Team, 1994.

[6] Jensen, K.: *Coloured Petri Nets: Basic Concepts, Analysis Methods and Practical Use*. Springer Verlag, Berlin, Heidelberg, New York, 1992.

[7] Perry, D.: *VHDL*. McGraw-Hill Inc., 1991.

[8] Peterson, J.: *Petri Net Theory and the Modelling of Systems*. Prentice-Hall Inc., 1981.

[9] *Visual HDL for VHDL on UNIX - Edition 4.0*. Summit Design Inc., 1997.

[10] Wikarski, D.: Petri net tools: A Comparative Study. ISST-Bericht 39/96. Technical Report, Fraunhofer-Gesellschaft e.G., Berlin, 1996.

Chapter 11

AN EVOLUTIONARY APPROACH TO THE USE OF PETRI NET BASED MODELS

From Parallel Controllers to HW/SW Codesign

Ricardo J. Machado, João M. Fernandes, António J. Esteves
and Henrique D. Santos

Dept. of Informatics, School of Engineering, University of Minho
4700-320 Braga, Portugal

rmac@dsi.uminho.pt

Abstract   The main purpose of this article is to present how Petri Nets (PNs) have been used for hardware design at our research laboratory. We describe the use of PN models to specify synchronous parallel controllers and how PN specifications can be extended to include the behavioural description of the data path, by using object-oriented concepts. Some hierarchical mechanisms which deal with the specification of complex digital systems are highlighted. It is described a design flow that includes, among others, the automatic generation of VHDL code to synthesize the control unit of the system. The use of PNs as part of a multiple-view model within an object-oriented methodology for hardware/software codesign is debated. The EDgAR-2 platform is considered as the reconfigurable target architecture for implementing the systems and its main characteristics are shown.

Keywords: Hardware Design, HW/SW Codesign, Petri Nets, Reconfigurable Architecture

Introduction

The control unit and the data path must both be considered by the specification model and the CAD environment, in order to fully specify a digital control system. The behaviour of the control unit of a digital system is usually described with a Finite State Machine (FSM). Whenever the system functionality presents concurrent activities, the specification

205

A. Yakovlev et al.(eds.), Hardware Design and Petri Nets, 205-222.
© 2000 *Kluwer Academic Publishers.*

of the system becomes more problematic and awkward, if we only use FSM-based techniques.

To specify a parallel controller using FSM techniques, there are, at least, two alternatives: (1) serially-linked controllers can be obtained by identifying sub-routines in the specification, or (2) concurrently-linked controllers should be connected with semaphore bits or common lines [24]. These solutions are generally awkward to apply, and can result in inefficient implementations due to the pre-partitioning, which limits the concurrency to the number of FSMs used. It is also hard to verify for parallel synchronization problems, such as deadlocks or multiple-sourcing.

Among the existing modelling paradigms, the PN-based one allows an easy specification of cooperative subsystems [23]. PNs are associated with a graphical notation, which is easy to understand and a system modelled with a PN might benefit from a mathematical theory to formally check its properties [27].

1.   PARALLEL CONTROLLERS

A development methodology for digital systems must provide tools for system specification, modelling and implementation. System specification includes the description of the expected behaviour (functionality) of the digital system. Modelling involves constructing a mathematical formalism embodying the specified system behaviour. This formalism can be manipulated and analysed to determine properties of the system which are not necessarily apparent from the initial problem statement, usually written in a natural language, such as english. The modelling formalism may also be adopted for the system specification, when there is a close correspondence between the system model and its specification.

Several researchers have shown that PNs are a powerful formalism to model the behaviour of parallel systems, namely, parallel digital controllers. A marking of the PN is equivalent to a global state of the modelled system (node in the reachability graph) and a change of the marking corresponds to a state transition (edge in the reachability graph). A detailed analysis of the model, based on a set of well established methods, allows the detection of a large number of design errors prior to the system implementation.

Several types of PNs were proposed to model digital systems, either by imposing restrictions to a basic formalism, or by adding extensions to it. PN-based controllers can be best modelled by safe PNs, which can be viewed as a natural extension to FSMs, providing an easy migration path from FSM to PN-based specifications. To effectively model parallel controllers using safe Place/Transition nets [23], the following modifications were proposed [6]: (1) logic expressions are assigned to transitions (the guards); (2) Moore type output signals are associated to places, while Mealy type output signals are related to transitions, to represent the controller actions; (3) transitions firing are synchronized with the active edge of a (global) clock; and (4) enabling and inhibitor arcs are supported.

The resulting PN type is called Synchronous Interpreted PN (SIPN), which can be used to specify the control unit of synchronous digital systems. Some PN formalisms, namely STGs (Signal Transition Graphs) [31], were also proposed to specify asynchronous digital circuits [30]. To execute an SIPN, all the enabled transitions at a given moment wait for a clock pulse and then all fire to produce a new marking. Fig. 11.1 presents an SIPN example, where xi are input signals and yi are output signals.

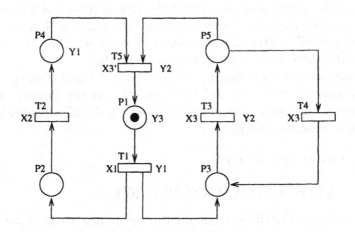

Figure 11.1 An SIPN specification of a controller.

A software framework was developed at our laboratory [6], to accept an SIPN-based controller specification, written in the CONPAR language, in order to validate the properties of the controller and to allow the PN model animation. Validation of models is important if they are to evolve into implementations, so a compiler was also included in the framework to generate the corresponding VHDL code [7]. This code can feed standard ECAD packages for simulation and synthesis purposes.

1.1   THE CONPAR DESCRIPTION LANGUAGE

PNs can also be viewed as formal models for logic rule-based specifications They make the straightforward link between algebraic numerical methods and the symbolic mathematical logic based methods of specification, optimisation, verification and synthesis. The rule-based form of specification can be considered as an alternative textual form of timing diagram description. The causality among signals is explicitly given in terms of local, relevant inputs, outputs and state changes.

The CONPAR description language, which is an extension to a previously defined language called PNSF [15], was developed to specify SIPN-based controllers, supporting macroplaces. In CONPAR notation, a transition is described as a conditional rule:

<label> : <PreConditions> |- <PostConditions> ;

The precondition and postcondition are respectively formed from input and output place symbols. When the preconditions of a rule are satisfied (hold), the postconditions are made true (they will hold). Logical conjunction of all related discrete states is assumed when the precondition contains more than one discrete state symbol.

For example, the transition t1 in Fig. 11.1 has input place p1, output places p2 and p3, it is guarded by input x1 and the output signal y1 is activated when the transition is enabled. In CONPAR notation, this transition is described as follows:

t1: p1 * x1 |- p2 * p3 * y1;

1.2   THE VHDL COMPILER

The CONPAR description language corresponds to an intermediate representation that links the SIPN model to the corresponding VHDL description. This transformation (from CONPAR to VHDL) can be obtained automatically by using a VHDL compiler already developed.

To obtain an efficient implementation, the PN is directly mapped into boolean equations without explicitly enumerating all possible global states and global state changes [1]. The specification is given in terms of the local states changes (local transitions) and one-hot code state assignment is used [24].

A VHDL textual PN specification of parallel controllers was proposed in [24], which describes a VHDL template with **ASSERT** statements to enable the syntatic and semantic correctness of the model to be tested. Experimental results, developed at Inmos in a practical design, achieved a 50% area reduction and a 40% speed improvement over the best FSM synthesis. In this work, this VHDL template was adopted for automatic code generation.

The VHDL code generated by our compiler is more readable than the one created with the CAMAD approach [25], where the VHDL code is not directly related to the original PN specification. This may cause some implementation inefficiency, since the PN is transformed into an FSM (which is built with the same algorithm as the reachability graph) and then translated into VHDL code using a CASE statement inside a PROCESS.

Examples on the use of SIPNs, the CONPAR language and the CAD environnment can be found in [10, 8, 9]. The complete grammar for the CONPAR is described in [7]

2.   PARALLEL CONTROLLERS + DATA PATH

The SIPN model was developed, aiming just the specification of the control part of the digital system: the data path of the system can not be described with the mechanisms available on the model. In some situations, this is considered to be a severe limitation, since it does not allow the integrated development of the whole hardware part of the system. For instance, the simulation task may become difficult because the information on the data path may have to be obtained from different simulation environments.

To overcome this limitation, the shobi-PN model [17], which is an extension to the SIPN model, was developed. The shobi-PN model supports hierarchy and allows objects to be used for specifying the data path resources. A full digital system can be specified and tested, following a structured and incremental approach.

The shobi-PN model presents the same characteristics as the SIPN model, in what concerns synchronism and interpretation, but adds new mechanisms by supporting object-oriented modelling ideas and new hierarchical constructs, in both the control unit and the data path. This model embodies concepts present in Synchronous PNs [3], Hierarchical PNs [5], Coloured PNs [14], and Object-Oriented PNs [16].

In the shobi-PN model, the tokens represent objects that model data path resources. The instance variables represent the information that is processed on the data path and the methods are the interface between the control unit and the data path. The tokens may be considered as coloured, if SIPN tokens are viewed as uncoloured (the SIPN places are safe). Each token models a structure of the data path.

A node (a transition or a place) invokes the tokens' methods, when the tokens arrive at that node. Nevertheless, only the methods that have a direct relation with the hardware control signals are directly invoked in the PN. There are additional methods available at the objects' interface that are not used by the PN. These methods are invoked by the simulation software to visualize the contents of a data path structure in any state of the PN.

Each arc is associated with one or more colours which indicate the types of objects that are allowed to pass through that arc. This means that, for each data path structure, there is a well-defined path on the PN. This restriction simplifies the PN and limits the capacity of some places, since it is not needed that objects, that are not invoked, unnecessarily traverse the PN.

Hierarchy can be introduced in the specifications in two different ways. The control unit is modelled by the PN structure, and to introduce the hierarchy on the controller, macronodes (representing sub-PNs) may be used. The data path resources are represented by the internal structure of the tokens, and the hierarchy can be introduced by *aggregation* (composition) of several objects inside one single token (a macrotoken) or by using the *inheritance* of methods and data structures.

2.1   SYNTHESIS OF THE CONTROLLER

For simulation purposes, the shobi-PN specification can be used directly, but to synthesize the control unit, the control part of a shobi-PN is transformed into an SIPN. This mapping is possible if it ensured that there is a structural compatibility in the control unit representation. Other topics, such as the PN reinitializations and the simultaneity on the invocation of different methods on the same node, are also important for the mapping but they are not considered in this paper: for details please refer to [17].

To ensure the structural compatibility of the control unit representation in the SIPN and shobi-PN models, it is imposed that the skeleton of the shobi-PN is structurally equivalent to a SIPN without reinitializations. The following concepts used for shobi-PNs are introduced: (1) Control Net: set of contiguous nodes and arcs of the shobi-PN that structurally corresponds to the SIPN without reinitilizations; (2) Control Track: path defined by a token in the Control Net; (3) Control Nodes: nodes (places or transitions) of the Control Net; (4) Control Arcs: arcs of the Control Net; (5) Closing Track: path defined by a token outside the Control Net; (6) Closing Nodes: nodes of a Closing Track; (7) Closing Arcs: arcs of a Closing Track; (8) Closing Cycle: path

defined by the movement of a token in the shobi-PN. It is composed by
a Control Track and also, if applicable, by a Closing Track. It can be
identified by the tracking of the colour associated with all the arcs of
the cycle; (9) Associated Net: SIPN structurally equivalent to the Control Net after the introduction of the reinitilizations for the uncoloured
tokens.

These concepts can be more easily understood by using the shobi-
PN in Fig. 11.2(a) to specify a simple control sequence, with two ob-
jects/tokens to model two structures of the data path.

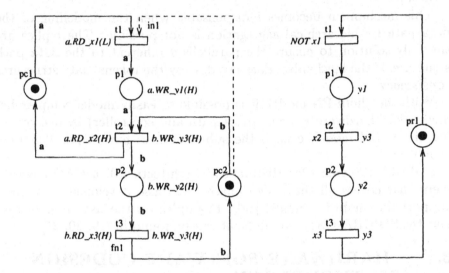

Figure 11.2   (a) shobi-PN for a simple control sequence and (b) its corresponding
SIPN.

In this example, the control net is composed by the following set of
nodes $\{t1, p1, t2, p2, t3\}$ and by the arcs that directly link them. It defines the skeleton of the shobi-PN. The control track for token a consists
of $\{t1, p1, t2\}$, while the control track for token b consists of $\{t2, p2, t3\}$.
The closing track for the token a consists of $\{t2, pc1, t1\}$ and for token b consists of $\{t3, pc2, t2\}$. The closing cycles for tokens a and b are
$\{t1, p1, t2, pc1\}$ and $\{t2, p2, t3, pc2\}$, respectively.

Mapping the shobi-PN into the associated net is made by transforming k-limited places into safe places. A safe place generates, whenever
marked, all the control signals associated to the methods invoked in the
corresponding place in the shobi-PN. As an example, consider the SIPN
in Fig. 11.2(b).

2.2   REPLICA MECHANISM

Whenever several methods that use the same data structures are concurrently invoked to a given token in different nodes, it is necessary to support a replica mechanism. This mechanism allows a token to be replicated as many times as needed, so that it is structurally possible to concurrently invoke methods to the same token, but in distinct areas of the PN. This mechanism can be used as an elegant solution for a complex problem (the multiple-sourcing) that could be alternatively, but innefficiently, solved at the algorithmic level, by changing the PN structure.

This mechanism becomes indispensable when the modelling of the data path by hierarchical aggregation is not possible. The replica are the only solution to ensure the parallelism inherent to the data path structure, if the mechanism does not destroy the tokens' data structures consistency.

With the shobi-PN model, it is possible to easily model complex behaviours of hardware systems (data path and controller) by decomposing the global model, even if the sub-structures have a parallel time-evolution.

SOFHIA (Software for Hierarchical Architectures), a CAD environment that covers all the design phases, was also developed to directly support the shobi-PN model [18]. Examples on the use of shobi-PNs and the SOFHIA CAD environment can be found in [19, 20, 21]

3.   HARDWARE/SOFTWARE CODESIGN METHODOLOGY

A methodology to system development based on the operational approach is essential to guarantee that complex systems can be addressed [32]. The main idea of this approach is based on an executable specification that evolves through transformational refinements to obtain the final implementation.

Object-oriented models (usually multiple view models covering the system's object, dynamic, and functional perspectives) are expected to fully address the above requirements, since they allow the easy refinement of application-domain objects during the whole process. However, there is no established methodology for hardware/software codesign that exploits the benefits of object-oriented system modelling techniques.

MOOSE is a graphical/textual method which is geared towards the development of embedded computer systems and leads to codesign after the system as a whole has been investigated through the use of abstract and executable models [22]. MOOSE uses a multiple view model for specifying the systems: OIDs (Object Interaction Diagrams) which are DFD-like diagrams for functional modelling, Domain Model for object modelling, and STD (State Transition Diagrams) to model the dynamic behaviour of the system.

Our methodology is based on MOOSE, with the following modifications: (1) STDs are replaced by shobi-PNs, which allow an easy handling of concurrency within the dynamic management of the system's objects; (2) the MOOSE paradigm follows essentially the waterfall process model, whilst we proposed a more iterative approach in the definition of the committed and the platform architectures, since we include the HDL/HLL generation in the partitioning phase (Fig. 11.3); (3) an umbrella testbed is included to cover all the development phases, to allow behavioural co-simulation, partitioning co-simulation and implementation co-verification; (4) MOOSE bases partitioning on the designer experience and intuition, while we use an automatic partitioning based on heuristics and an iterative refinement through resources area and time estimation; and (5) a target architecture (EDgAR-2) is used for the hardware parts implementation.

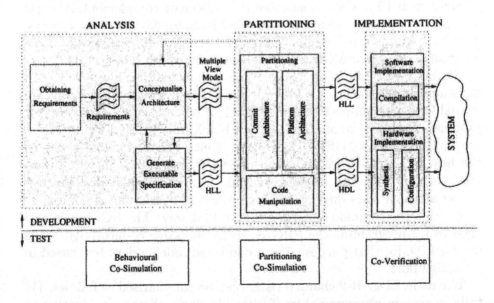

Figure 11.3 The proposed methodology.

The parallel capabilities of PNs are essentially used during the partitioning activities, which refine the executable specifications towards an equivalent CFSMD model (specified by an HDL) to map it into the hardware reconfigurable components (section 4.).

4.   EDGAR-2: THE RECONFIGURABLE TARGET ARCHITECTURE

4.1   THE ARCHITECTURE

EDgAR-2 is an FPGA/CPLD based system used for hardware/software codesign and rapid system prototyping. The EDgAR-2 is the successor of the EDgAR. Both systems were developed at our laboratory. The EDgAR was first conceived as part of a stand alone emulation tool for digital systems [4]. The EDgAR-2 is an enhancement architecture, including updated and powerful devices, with In System Programmable (ISP) property and a PCI bus interface, which allow it to be a reconfigurable hardware block in a host PC, implementing a codesign machine or even a high versatile prototype tool for digital design.

Programmable logic devices (PLDs) can be divided into two classes: one based on coarse grain two level logic blocks, with guaranteed time delay (CPLDs), typically used for control paths or time critical circuits; the other based on fine grain multi level logic blocks (FPGAs), typically used for data paths or space critical circuits [26].

Since each PLD class is suitable to implement complementary parts in a typical digital system, the EDgAR-2 includes devices of both types. The basic architecture element (Fig. 11.4) is a module composed of an array of 4 Processor Modules (PMs), including each one a control unit and a data path unit. The PMs are interconnected in a linear way with dedicated buses, forming a PM pipeline. Both sides of the array are available to interconnect several EDgAR-2 boards, in a larger array or pipeline. Each PM is implemented with a Xilinx 4010E FPGA [28] — data path — and a 211SP MACH [2] — control path. In what concerns the host PC, each PM is linked to one byte from the 32 bits PCI data bus. In this way, the software module in a codesign realisation can access all the 4 PMs during the same bus cycle, assuming it is possible to manage the common address space in that way. The PCI bus is also used to (re)programme the FPGAs, using the same connectivity, while the CPLDs, for that purpose, use a dedicated independent bus based in a parallel port.

The main EDgAR-2 characteristics can be summarised as follows: (1) fully in system programmable; (2) flexible clock schemes (respecting to frequency and source), with a limited support for asynchronous problems; (3) polling and interrupt mechanism for communication with the host system; (4) PCI burst mode support; (5) pipeline structure; and (6) scalable architecture.

Because EDgAR-2 is full in system programmable, it supports the reconfigurable paradigm. Together with a real time operating system

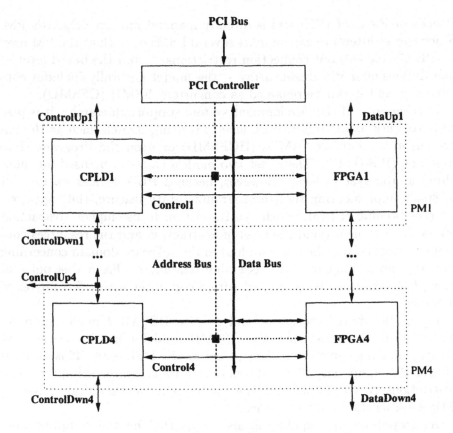

Figure 11.4 The EDgAR-2 architecture.

(RTOS) and a custom PCI based computer architecture, the obtained machine is a powerful tool to solve time critical problems. However, during the initial phase and for validation purposes, the prototype operates only under control of the Windows NT (tm) operating system, using a proprietary device driver.

4.2   THE COMPUTATIONAL MODEL

The EdgAR-2 architecture was designed to directly accommodate a finite state machine with data path (FSMD) model [13]. This model is basically an extension to the well known FSM model, with a data path to support a higher level of data abstraction, including primitive variable objects and the associated logical and arithmetic operators. From the structural point of view, and taking into account the respective properties, the FSMD model is composed of 2 components: an FSM controller and a data path. This (FSM controller, data path) pair can be called a

Processor Element (PE) and is directly mapped into an EDgAR's PM. Since the architecture can support several FSMDs, both at the PM level — PE cluster without connection restrictions — and the board level — one dimensional PE cluster array -, the model naturally includes concurrency, and it can be renamed as concurrent FSMD (CFSMD).

Furthermore, if the development system supports hierarchy, it is possible to work at an architecture level [11] using models such as the Hierarchical Concurrent FSMDs (HCFSMD) or even the Program State Machine (PSM) [12]. These last two models can also be used for modelling at the system level. However, because PLD devices waste a lot of space implementing the (re)programmability feature, the logical resources become critical, which is a restriction to use higher abstraction levels. This issue around the level of abstraction and the resources consumed is essentially the same we find in the software domain concerning the high level languages vs. assembly languages. From this point of view, EDgAR-2 can be compared with a computer with a few kbytes of memory.

From the above discussion, and despite EDgAR-2 model supports higher level of abstractions, the CFSMD model seems to be the best choice, allowing concurrent descriptions at the RTL level. This decision affects the complexity of the development tools, and can impose some restrictions to the implementation of codesign methodologies, having EDgAR-2 as the target hardware.

As stated before, pipeline is also supported by the computational model of the architecture. At the outer level EDgAR-2 can be defined as a pipeline of PE clusters, with each cluster formed by arbitrary concurrent PEs. This architecture is well suited, for instance, to a production line with several interdependent machines, each with a set of concurrent processes. At the PE level pipeline is not explicit in the architecture, but especially the FPGAs easily allow the addition of pipeline stages to the data path. To what concerns the control path, typically it is not profitable to introduce pipeline stages, due to the faster operation of the 2-level logic CPLDs.

To summarise, the target hardware defines an elementary PE architecture, comprising: (1) an FSM controller block (CPLDs) — supporting registers and two-level logic. This block does not include enough memory to be micro-programmed. The micro-programmability would allow a microprocessor model, which is not needed because the host has a processor. (2) a data path block (FPGAs) — supporting logic and arithmetic operators up to 32 bits wide, data structures with low complexity (up to 4 * 1kbyte) and the common logic structures (MUXs, decoders, registers, ALUs up to 32). A system is composed of an arbitrary number of concurrent PEs, forming clusters, eventually linked in a pipeline structure.

4.3 THE COMMUNICATION MODEL

At a high level of abstraction two communication class mechanisms can be identified: message passing and shared memory. Some of the operations typically found in a message passing mechanism are: point-to-point communication (one to one relation), broadcasting (one to all relation), scatter (one node sends a distinct messages for each node), gather (one node gets a distinct messages from each node) [29]. Shared memory evolves the coordinated access of all processes to a common memory space, requiring the definition of arbitration rules. EDgAR-2 does not include a large common memory, and so it imposes severe restrictions to the implementation of shared memory mechanisms.

At a lower level of abstraction, a physical topology imposes more or less restrictions to the implementation of the above communication mechanisms. Some of the topologies commonly used on systems like the EDgAR-2 are: (1) chain — all nodes, except the two end nodes, communicates with two neighbours, building a chain; (2) ring — a chain topology, where the two end nodes are connected; (3) n-dimensional mesh — every node communicates with its adjacent ones, building a n-dimensional cube; (4) centralised or star — a topology where one single node (concentrator) connects to all the others; (5) hierarchical or tree — the connections between nodes present an hierarchical or tree-like structure; (6) complete or fully connected topology — every node communicates to all the others; and (7) an irregular topology.

These topologies can be explicitly defined in the architecture design, or implemented in software or even trough programmable routing devices, allowing topology changes.

To what concerns EDgAR-2, it is necessary to distinguish the communication between processes running on hardware, on software and on both hardware and software. The processes running on hardware must conform to the CFSMD computational model proposed. If this model is not restricted, it requires a fully connected topology at the PE level. Both the CPLDs and FPGAs programmable structures are very rich in terms of in chip interconnections, allowing, for most applications, a fully connected topology at the PE cluster level. To implement the same topology at the PM level it is necessary to use larger chips with much more pins, which imposes a higher cost and a higher complexity of the PCB routing.

The analysis of the CFSMD model for several problems — (embedded) control systems domain — showed that only a small number of systems will justify a fully connected topology. So it was decided to provide the architecture of the EDgAR-2 with the following combined topology: a fully connected or centralised topology at the PE cluster level, with a chain (or optionally a ring) topology at the PM or system level.

The processes running on software naturally use the mechanisms supported by the host operating system. This issue has no influence in the EDgAR-2's model and will not be more detailed here. Finally, in a codesign environment, there will be hardware processes communicating with software processes. As described before, the host can access all the PMs through a PCI bus. For complexity reasons and because typically EDgAR-2 works as a coprocessor, the PCI interface does not implement Bus Master operations. This decision implicitly give to the software processes the control over hardware processes. This corresponds to the centralised mechanism defined above. Fig. 11.5 shows the communication model just described.

Any communication involving EDgAR-2 can be synchronous or asynchronous. The later requires some type of handshake, while the former just requires a signal from the sender to the receiver. Besides, there will be some type of protocol, which is naturally limited by the amount of resources available. Typically, a communication should be restricted to a register transfer with a request/acknowledge handshake. If a more elaborated protocol and/or handshake is required, its specification must be described, making it in an additional PE.

Using the communication topology described, higher level communication models can be implemented. The decision of which model to use will be done by the development methodology (however, it is not difficult to realise a complex communication protocol which could waste most of the EDgAR-2 resources). Some guidelines for this decision are described here. First, communication models based on large memory utilisation are not supported due to the lack of memory on the EDgAR-2 board. Second, the broadcasting operation used on the message-passing model is not suitably supported to other levels rather than PE cluster level, because it will require the architecture to allow simultaneous write operations to all devices, which is not the case — to execute a message broadcasting a time overhead is imposed by the required sequence of point-to-point communications. Third, it is necessary to use low complexity communication protocols, in order to achieve a good balance between communication resources and processing resources.

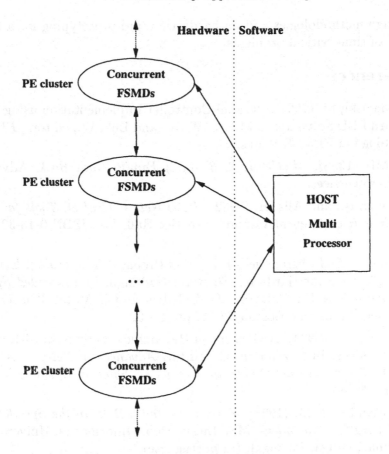

Figure 11.5 The EDgAR-2 communication model.

5.   CONCLUSIONS

This paper has presented the evolution of a family of PN-based models to allow the specification of sequential and parallel controllers, with or without data path, and has also shown how is it possible to use them within a wider methodology for hardware/software codesign. PNs are viewed as a fundamental component of a multiple-view model to deal with parallel and concurrent behavioural specification at system-level design. The whole methodology is object-oriented and follows the operational approach to allow the generation of executable specifications and transformational refinements to obtain the hardware and software solutions for system implementation. The codesign target architecture is also presented, which includes a CPLD/FPGA-based ISP board and a host PC with real-time multiprocessing capabilities. The target architecture proposed offers an interesting low cost environment to research on codesign methodologies and on hardware rapid-prototyping for a broad range of time critical problems.

References

[1] Adamski, M. (1991). Parallel Controller Implementation using Standard PLD Software. In Moore, W. R., and Luk, W., editors, *FPGAs*. Abingdon EE&CS Books.

[2] AMD (1995). *MACH 1, 2, 3 and 4 Family Data Book*. Advanced Micro Devices.

[3] David, R., and Alla, H. (1992). *Petri Nets & Grafcet; Tools for modelling discrete event systems*. Prentice-Hall, UK. ISBN 0-13-327537-X.

[4] Esteves, A. J., Fernandes, J. M., and Proença, A. J. (1997). EDgAR: A Platform for Hardware/Software Codesign. In *Embedded System Applications*, Ed. C. Baron, J.-C. Geffroy and G. Motet, Kluwer Academic Publishers, Boston, USA, pp. 19–32.

[5] Fehling, R. (1993). A Concept of Hierarchical Petri Nets with Building Blocks. In Rozenberg, G., editor, *Advances in Petri Nets 1993*, vol. 674 of *Lecture Notes in Computer Science*, Springer-Verlag, pp. 148–68.

[6] Fernandes, J. M. (1994). *Petri Nets and VHDL in the Specification of Parallel Controllers*. MSc thesis, Dept. Informatics, University of Minho, Braga, Portugal. (in portuguese)

[7] Fernandes, J. M., Adamski, M., and Proença, A. J. (1997). VHDL Generation from Hierarchical Petri Net Specifications of Parallel Controller. *IEE Proceedings: Computers and Digital Techniques*, 144:127–37.

[8] Fernandes, J. M., Pina, A. M., and Proença, A. J. (1995a). Concurrent Execution of Petri Nets based on Agents. In *1st Workshop on Object-Oriented Programming and Models of Concurrency within the XVI International Conference on Applications and Theory of Petri Nets*, Torino, Italy.

[9] Fernandes, J. M., Pina, A. M., and Proença, A. J. (1995b). Simulation and Synthesis of Parallel Controllers based on Petri Nets. In *VII Simpósio Brasileiro de Arquitetura de Computadores — Processamento de Alto Desempenho (SBAC-PAD'95)*, pp. 481–92, Canela, Brazil. (in Portuguese).

[10] Fernandes, J. M., and Proença, A. J. (1994). Petri Nets in the Specification and Validation of Parallel Controllers. In *1o. Encontro Na-*

cional do Colégio de Engenharia Electrotécnica, pp. 113–8, Ordem dos Engenheiros, Lisboa, Portugal. (in portuguese).

[11] Gajski, D. and Kuhn, R. (1983). Guest's editors introduction: New VLSI Tools. *IEEE Computer*, 16:11–4.

[12] Gajski, D., Marchioro, G., and Zhu, J. (1997). *Essential Issues in Codesign*, pp. 1–45. Kluwer Academic Publishers.

[13] Gajski, D. D., Dutt, N. D., and Wu, A. C.-H. (1994). *High-Level Synthesis: Introduction to Chip and System Design.* Kluwer Academic Publishers, 3 edition.

[14] Jensen, K. (1992). *Coloured Petri Nets: Basic Concepts, Analysis Methods and Practical Use*, vol. I. Springer-Verlag, Berlin, Germany.

[15] Kozlowski, T. (1993). *Petri-net-based CAD tools for parallel controller synthesis.* MSc thesis, University of Bristol, England.

[16] Lakos, C. (1995). The Object Orientation in Object Petri Nets. In *1st Workshop on Object-Oriented Programming and Models of Concurrency within the 16th International Conference on Applications and Theory of Petri Nets*, Torino, Italy.

[17] Machado, R. J. (1996). *Hierarchy in Object-Oriented Petri Nets for the Specification of Digital Systems.* MSc thesis, Dep. Informática, Universidade do Minho, Braga, Portugal. (in Portuguese).

[18] Machado, R. J., Fernandes, J. M., and Proença, A. J. (1997a). SOFHIA: A CAD Environment to Design Digital Control Systems. In Kloos, C. D. and Cerny, E., editors, *XIII IFIP Conference on Computer Hardware Description Languages and Their Applications (CHDL'97)*, pp. 86–8, Toledo, Spain. Chapman & Hall.

[19] Machado, R. J., Fernandes, J. M., and Proença, A. J. (1997b). Specification of Industrial Digital Controllers with Object-Oriented Petri Nets. In *IEEE International Symposium on Industrial Electronics (ISIE'97)*, vol. 1, pp. 78–83, Guimarães, Portugal.

[20] Machado, R. J., Fernandes, J. M., and Proença, A. J. (1998). An Object-Oriented Model for Rapid Prototyping of Data Path/Control Systems — A Case Study. In *9th IFAC Symposium on Information Control in Manufacturing (INCOM'98)*, vol. 2, pp. 269–74, Nancy and Metz, France.

[21] Machado, R. J., Fernandes, J. M., and Proença, A. J. (1998). Hierarchical Mechanisms for High-level Modelling and Simulation of Digital Systems. In *5th IEEE International Conference on Electronics, Circuits and Systems (ICECS'98)*, vol. 3, pp. 229–32, Lisbon, Portugal.

[22] Morris, D., Evans, G., Green, P., and Theaker, C. (1996). *Object-Oriented Computer Systems Engineering.* Applied Computing. Springer-Verlag, London, U.K. ISBN 3-540-76020-2.

[23] Murata, T. (1989). Petri Nets: Properties, Analysis and Applications. *Proceedings of the IEEE*, 77(4):541–80.

[24] Pardey, J. and Bolton, M. (1991). Logic Synthesis of Synchronous Parallel Controllers. *Proceedings of the IEEE International Conference on Computer Design*, pp. 454–7.

[25] Peng, Z. (1992). Digital System Simulation with VHDL in a High-level Synthesis System. *Microprocessing and Microprogramming*, 35:263–70.

[26] Santos, H. D. (1996). *Specification and Analysis Methodologies of Digital Systems: Development of an APA (GLiTCH) controller.* PhD thesis, Departamento de Informática, Universidade do Minho, Braga, Portugal. (in portuguese).

[27] Valette, R., Courvoisier, M., Bigou, J., and Albukerque, J. (1983). A Petri Net Based Programmable Logic Controller. In *IFIP First International Conference on Computer Applications in Production and Engineering.*

[28] Xilinx (1996). *The Programmable Logic Data Book.* Xilinx.

[29] Xu, Z. and Hwang, K. (1996). Modelling Communication Overhead: MPI and MPL Performance on the IBM SP2. *IEEE Parallel & Distributed Technology*, pp. 9–23.

[30] Yakovlev, A., Koelmans, A., Semenov, A., and Kinniment, D. (1996a). Modelling, Analysis and Synthesis of Asynchronous Control Circuits Using Petri Nets. *INTEGRATION: the VLSI Journal*, (21):143–70.

[31] Yakovlev, A., Lavagno, L., and Sangiovanni-Vincentelli, A. (1996b). A unified signal transition graph model for asynchronous control circuit synthesis. *Formal Methods in System Design*, (9):139–188.

[32] Zave, P. (1984). The Operational versus the Conventional Approach to Software Development. *Communications of the ACM*, 27(2):104–18.

Chapter 12

MODELLING AND IMPLEMENTATION OF PETRI NETS USING VHDL

Dave Prothero
*Centre for Concurrent Systems and VLSI ***
South Bank University
Borough Road, London SE1 0AA, UK
prothed@sbu.ac.uk

Abstract   The use of Petri nets for the modelling of concurrent systems is long-established. Their application in the field of digital design is to the specification and modelling of self-timed systems within which the sequencing of operations is not synchronised to a global clock signal. This paper demonstrates the use of a standard hardware description language, VHDL, to model and simulate the operation of Petri nets at different levels of abstraction. Techniques for subsequently synthesising hardware circuits from Petri net specifications are also described.

Keywords: asynchronous, HDL, modelling, Petri nets, synthesis

1.   INTRODUCTION

This paper introduces the use of the VHDL hardware description language to model the structure and behavior of place-transition nets. A large number of computer-based tools have been developed to support the construction, animation (simulation) or analysis of simple and extended Petri nets, (currently, over 35 such tools are listed at [13]), however, such tools typically use individual descriptive techniques or languages so that it is not possible to transport a given Petri net description between tools. This problem may be eliminated with the use of a standard language such as VHDL.

* This work is supported by the European Commission under Working Group 21949 ACiD-WG as part of the Esprit Fourth Framework.

A. Yakovlev et al.(eds.), Hardware Design and Petri Nets, 223-236.
© 2000 *Kluwer Academic Publishers.*

The "Very high speed integrated circuit Hardware Description Language" was originally developed in the early 1980's to allow vendor-independent descriptions of digital electronic components and systems. VHDL is now formally defined by IEEE Standard 1076-93 and has become a world-wide standard in the electronics industry. VHDL descriptions may be easily transported, allowing data interchange between groups of designers. The semantic interpretation of VHDL operations is defined in the context of event-driven simulation, rather than in continuous time, which corresponds naturally to the interpretation of Petri net operations.

In practice, VHDL has been much more widely used for the 'hardware level' modelling of digital systems than for the modelling of abstract, uninterpreted structures such as Petri nets. However, it will be shown that VHDL can provide the facilities required for both the simulation and analysis (such as performance modelling and correctness checking) of both simple and extended Petri nets. Furthermore, since VHDL can be used to model systems at any level of abstraction, the same language can be used throughout the design process from an initial specification to a hardware specification. This paper will present examples of Petri nets modelled at different levels of abstraction, namely -

- An abstract model in which only transitions are explicitly described, places being implicit within the arcs between transitions.
- A behavioral model in which both transitions and places are described, allowing greater flexibility in the specification of their behavior.
- A hardware model in which transitions and places are implemented as digital circuits. The resulting system then operating as a self-timed controller suitable for use in the conventional controller/datapath paradigm for complex digital systems [15] [21].

In the examples presented, several (hopefully intuitive) VHDL constructs are used without detailed explanation. For a full introduction to VHDL, several texts are available including [1, 10, 11, 4].

2.   DEFINITIONS

In order to discuss the construction and behavior of Petri net models, some basic definitions must be given. These are closely styled upon those described by Peterson [12] and are consistent with common practice.

A Petri net may be identified as a bipartite, directed graph $N = (T,P,A)$

where: $T = \{t_1, t_2, t_3 \ldots t_n\}$ is a set of *transitions*, $P = \{p_1, p_2, p_3 \ldots p_n\}$ is a set of *places*, and $A \subseteq \{T \times P\} \cup \{P \times T\}$ is a set of directed *arcs*. Collectively, the sets of transitions and places constitute the nodes of N, with A defining the node interconnection. A marking M of a Petri net is a pattern of tokens occupying places and represents the state of the system. M can be viewed as a vector whose i^{th} component M_i represents the number of tokens M assigns to p_i. A Petri net N = (T,P,A) with marking M is a 'marked Petri net' K = (T,P,A,M).

A transition in a marked Petri net is *enabled* if each of its input places contains at least one token. An enabled transition can fire by removing a token from each of its input places and putting a token in each output place. A token can be removed from a place by only one transition at a time.

Note that this abstract model is uninterpreted, meaning that no meaning is ascribed to the actions represented by the marking of places or the firing of transitions - other than that which may be suggested to the reader by their labeling.

Many extensions to the simple Petri net model have been defined in order to increase their modelling power and decision power. For example - weighted nets, timed nets, coloured nets, interpreted nets, queuing nets, etc. These extensions are not discussed here but examples are introduced later in order to illustrate the use of VHDL to cope with such extensions.

Petri nets have been widely used as a basis for the analysis of various properties of systems, especially those containing concurrency. For example - the verification of properties such as liveness and safety in uninterpreted models, the verification of determinacy and deadlock-freedom in interpreted models and performance modelling such as the estimation of throughput and latency time in timed models [7, 6].

3. MODELLING WITH VHDL

The use of hardware description languages such as VHDL for the modelling of digital circuits at the register-transfer and gate levels (ie, at a detailed level of interpretation) is widespread, however, relatively little published work refers to the use of VHDL at higher levels of abstraction. In this section, a brief introduction to VHDL and its application to Petri net modelling will be given.

As in any hardware description language, VHDL provides for the description of components and the connections between them. What distinguishes VHDL from many other languages is the ability to define new data types and the operations upon them, thereby allowing the definition of a *package* of types, objects and functions to support the modelling of any required type of system. The behavioral semantics of VHDL are defined in terms of a discrete-time model corresponding to the operation of an event-driven simulator. However, Petri net firing rules are consistent with such a timing model and so this presents no limitation.

In VHDL, each component is represented by a description consisting of two parts - an *entity*, which describes the external connections and generic parameters of the component, and an *architecture*, which describes the internal structure or behavior of the component by use of statements similar in appearance to those found in high-level programming languages. Also, an architecture may contain instantiations of other VHDL descriptions, directly supporting hierarchical descriptions. A significant difference between VHDL and conventional (procedural) languages is that VHDL statements are generally evaluated concurrently, not sequentially. In the case of Example 1, the outputs **sum** and **difference** will be evaluated whenever the input signals change value and will both change value at the same point in time.

```
entity alu is                                      -- simple ALU
  port (a, b : in integer;  sum, difference : out integer);
end alu;

architecture functional of alu is    -- description of behavior
begin
  sum <= a + b ;
  difference <= a - b ;
end functional;
```

Example 1

3.1   Petri net modelling with implicit places

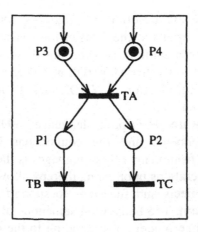

Figure 12.1. Simple timed Petri net

In this case, Petri net transitions will be modelled by VHDL block statements, which are evaluated (executed) when the guard expression following the **block** keyword is true. The conditions within the guard expression , for example –

block(input_place='1' and output_place='0') then correspond to the Petri net firing conditions. The places are represented by the signals between the transitions and are declared as **register** types, allowing them to retain the values assigned to them by their associated transitions.

Figure 12.1 shows a simple timed Petri net (a marked graph or state machine) and Example 2 a corresponding VHDL description.

```
use work.std_logic_1164.all;

-- Simple timed Petri Net
entity net is
end net;            -- this design has no external signals

architecture abstract of net is
  -- signals between transitions represent net places
  signal p1 , p2 : std_logic register :='0';
  signal p3 , p4 : std_logic register :='1';
                 -- note assignment of initial values

begin

TA: block (p3='1' and p4='1')        -- firing conditions
    begin
      p3 <= guarded '0' after 3 ns;
      p4 <= guarded '0' after 3 ns;
      p2 <= guarded '1' after 3 ns;
      p1 <= guarded '1' after 3 ns;
    end block;

TB: block (p1='1')
    begin
      p1 <= guarded '0' after 6 ns;
      p3 <= guarded '1' after 6 ns;
    end block;

TC: block (p2='1')
    begin
      p2 <= guarded '0' after 10 ns;
      p4 <= guarded '1' after 10 ns;
    end block;

end abstract;
```

Example 2

This model is uninterpreted, that is, the firing of the transitions does not correspond to any operation other than the assignment of token values to places. Also, this VHDL description does not contain any form of in-built verification or error-checking, for instance, that a place may contain at most one token. Such checks on signal values can be added by use of a *resolution function* which is called automatically when any component assigns a new value to that signal. (In Example 2, the package which defines the `std_logic` type also contains the resolution function). In Example 3, the user-defined function `resolved_place` detects an attempt by more than one source to assign a token to a place which is declared using a `bit` type and can consequently only adopt the values zero and one.

```
function resolved_place (drivers : bit_vector) return bit is
 variable temp : integer := 0;
begin
  for n in drivers'range loop
  if drivers(n)='1' then temp := temp+1; end if;
  end loop;
  assert not(temp > 1) report "Multiple drivers active for
place"
        severity warning;
  if temp=0 then return '0'; else return '1'; end if;
end resolved_place;
```

Example 3

Note that the number of drivers is evaluated dynamically each time the function is called. The assert statement executes if its argument is false, ie, if the assertion fails.

The resolution function must then be associated with a new signal type, then allowing signals to be declared, eg.

```
subtype place is resolved_place bit;
signal p5 , p6 : place ;
```

A further example of this style of modelling is given in [2] (although the timed transition model presented there (section 5.3.1.4) appears to contain a mistake which has been corrected in the descriptions shown in Example 2 above).

It can also be seen that the use of integer or enumerated signal types to represent places can allow weighted or coloured nets to be modelled.

Checks involving the correct behavior of transitions can similarly be incorporated by adding the appropriate code to the architecture of the component representing the transition.

3.2   Place-Transition net modelling

This section illustrates the explicit modelling of both places and transitions. Example 4 shows a VHDL description of a place capable of holding one token (similar in effect to an edge-triggered bistable).

```
use work.std_logic_1164.all;

entity place is
 generic (initial_token : integer := 0);
 port(s, r : in std_logic; q : out std_logic);
end place;

architecture rtl of place is
 begin
  process (s, r)   -- process activated when event occurs
                   -- on either signal
    begin
     if now = 0 ns then   -- test current simulation time
       if initial_token > 0 then q <= '1';
                            else q <= '0';
       end if;
     end if;
     if    s='1' and s'event then q <= '1';
                            -- normal operation
     elsif  r='1' and r'event then q <= '0';
     end if;
   end process;
 end rtl;
```

Example 4

Note the use of a generic parameter **initial_token** which is used to set the initial state of the place.

Example 5 shows another description of the Figure 12.1 Petri net. In this case, the transitions have been written using VHDL *processes*, inside which the enclosed statements are evaluated sequentially. The processes themselves are concurrent VHDL statements (as were the blocks in the previous example) and again provide a natural means of describing Petri net components.

```
use work.std_logic_1164.all;

entity net2 is
end net2;

architecture abstract of net2 is

 component place
  generic (initial_token : integer := 0 );
```

```
   port(s,r : in std_logic; q : out std_logic);
 end component;

 signal p1q , p2q , p3q , p4q  : std_logic ;   -- place outputs
 signal sp1 , sp2 , sp3 , sp4  : std_logic ;   -- set signals
 signal rp1 , rp2 , rp3 , rp4  : std_logic ;   -- reset signals


begin

P1: place port map (sp1 , rp1 , p1q);     -- instances of places
P2: place port map (sp2 , rp2 , p2q);
    -- P3 and P4 contain initial tokens
P3: place generic map (1) port map (sp3 , rp3 , p3q);
P4: place generic map (1) port map (sp4 , rp4 , p4q);

TA: process            -- transitions represented by processes
    begin
      -- wait until input places active
      wait until p3q='1' and p4q='1';
      assert false report "TA enabled" severity note;
      wait for 3 ns;      -- 'wait' represents transition time
      -- reset input places, set output places
      rp3 <='1','0' after 1 ns; rp4 <='1','0' after 1 ns;
      sp1 <='1','0' after 1 ns; sp2 <='1','0' after 1 ns;
      assert false report "TA terminated" severity note;
    end process;

TB: process
    begin
      wait until p1q='1';
      assert false report "TB enabled" severity note;
      wait for 5 ns;
      assert false report "TB terminated" severity note;
      rp1 <= '1','0' after 1 ns;
      sp3 <= '1','0' after 1 ns;
    end process;

TC: process
    begin
      wait until p2q='1';
      assert false report "TC enabled" severity note;
      wait for 8 ns;
      rp2 <= '1','0' after 1 ns;
      sp4 <= '1','0' after 1 ns;
      assert false report "TC terminated" severity note;
    end process;

end abstract;
```

Example 5

Note the use of explicit signals sp1, rp1 etc, to place and remove tokens, ie, to set and reset the bistables representing the places.

As in the example of the previous section, the place and transition models may be elaborated in order to extend the analytical power of the model. For example, reference [9] describes in detail a model of this general form which includes a number of correctness checks and is also extended to include a queuing capability in order to evaluate performance measures. Also, work such as that reported in [3] on the formal verification of generalised VHDL descriptions may be applied.

3.3   Hardware interpretation of Petri net components

In this section, the operations of the place and transition components are implemented by means of logic equations - corresponding to a logic circuit which has been synthesised from a Petri net specification. In one sense, the direct hardware implementation of an uninterpreted Petri net is a simple syntax-directed translation task since the specification language consists only of two well-defined components. In practice however, the actions implied by the firing of a transition must be interpreted as some form of meaningful operation, typically corresponding to the activation of a datapath component within the hardware system.

The final example (an elaboration of the place-transition net above) indicates how hardware circuit designs may be described using VHDL.

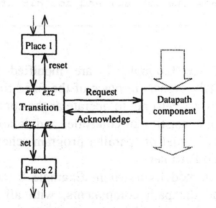

Figure 12.2. Fragment of a design

Figure 12.2 shows a fragment of a design in which the firing of a transition corresponds to the activation of a datapath component. Here, the place component, shown in Example 6, is described by simple Boolean functions. If required, specific hardware components can be instantiated directly within VHDL, but it is now commonplace to use a technology-independent representation which is then mapped into a particular library of components by a logic synthesis package such as SIS [18] or by a manufacturer-specific tool targeted at CPLD or FPGA devices [16, 19].

```
entity place is
  port (s, r, clr : in bit;  q : out bit);
end place;

architecture eqns of place is
  signal sa, qb : bit;
  begin
    sa <= s and not clr;    -- clr is a global hardware reset
    qb <= not sa and (r or qb or clr) after 1 ns;
    q <= not qb;                             -- place output
  end eqns;
```

Example 6

```
entity transition is
  port (ad, ex, ez, clr : in bit;  exz, rd : out bit);
end transition;

architecture eqns of transition is
  signal s3, csc : bit;
  begin
    rd <= ex and not ez and not csc after 1 ns;
    exz <= csc and not ad after 1 ns;
    s3 <= ez and not ex after 1 ns;
    csc <= not s3 and (ad or csc) and not clr after 1 ns;
  end eqns;
```

Example 7

The transitions, see Example 7, are modified to include request-acknowledge handshake signals in place of the fixed time delays previously included. This allows the Petri net to operate as the controller in a self-timed asynchronous system consisting of controller and datapath - based upon the classic computational model of 'parallel program schemata' [8], which are essentially interpreted Petri nets.

The complete net model, shown in Example 8, contains instances of place, transition and datapath components, with all VHDL descriptions consisting only of Boolean equations.

```
entity dcomp is                      -- datapath component
  port(req : in bit; ack : out bit);
end dcomp;

architecture abstract of dcomp is
  begin
    ack <= req after 25 ns;
  end abstract;
```

```
-- Hardware net description -------------------------------

entity net is
  port (reset, run : in bit);              -- external signals
end net;

architecture hardware of net is

  component place                      -- component declarations
    port (set, reset, clr : in bit;  q : out bit);
  end component;

  component transition
  port (ack, ex, ez, clr : in bit;  exz, req : out bit);
  end component;

  component dcomp
  port(req : in bit; ack : out bit);
  end component;

  signal pp, ps, rs, rd, ad  : bit ; -- signals between
                                     -- components

begin                          -- component instantiations

P1: place port map (run, rs, reset, pp);
P2: place port map (rs, run, reset, ps);

TA: transition port map (ad, pp, ps, reset, rs, rd);

DP: dcomp port map (rd, ad);

end hardware;
```

Example 8

A complete design implemented in this way will consist of a controller, whose temporal behavior is described by the original Petri net and which will consist of a network of place and transition components whose structure is isomorphic to the Petri net, plus a set of datapath components, each of which is activated by one or more transitions. Further design details of the controller and datapath components used in such a system are given in [14]. Such a direct translation of a Petri net specification to a controller circuit may be sub-optimal in the sense of logic synthesis, since each place and transition component will contain a bistable circuit and the total number of bistables will be proportional to the number of states in the original net. If the net is considered as a whole, rather than element-by-element, then a single asynchronous logic circuit with n states and $\log_2 n$ bistables may be derived. It is generally not feasible to design asynchronous logic circuits containing more than a few states by hand, and CAD tools, notably Petrify

[5] are used to derive the circuit equations. As a simple example, the equations for the transition component of Example 7 were generated from the Petri net shown in Figure 12.3 using Petrify.

Other approaches to the hardware synthesis problem include the Sutherland 'micropipeline' [20] which has found wide acceptance, see for example [17].

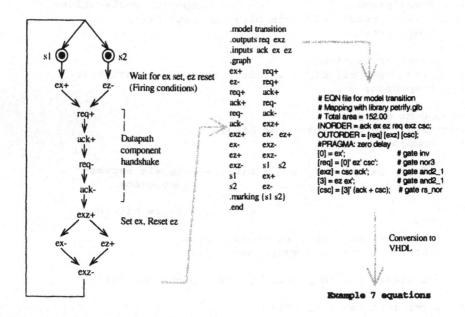

Figure 12.3. Synthesis example

4.   CONCLUSION

This paper has shown that VHDL may be used to model Petri nets at a range of abstraction levels. In practice, a net model may contain any mixture of these levels, allowing hierarchical or multilevel descriptions. Analytical techniques, both aimed at formal verification and performance estimation, have been reported. The fact that VHDL can support mixed-level descriptions and verification within a single multi-platform language recommends its wider adoption by the Petri net community.

5.   REFERENCES AND BIBLIOGRAPHY

[1]   Ashenden PJ. The designers guide to VHDL. Morgan Kaufmann. 1996.

[2]   Berge JM, et al. VHDL designers reference. Kluwer. 1991.

[3]   Borrione D, et al. Formal verification of VHDL descriptions in the Prevail environment. IEEE Design & Test of Computers. June 1992. pp. 42-55.

[4]   Coelho DR. The VHDL Handbook. Kluwer. 1989.

[5]   Cortadella J, et al. Petrify home page –
http://www.ac.upc.es/vlsi/petrify/petrify.html
Universitat Politecnica de Catalunya. 1997.

[6]   Foulk PW. CAD of concurrent computers. Research Studies Press. 1985.

[7]   Holiday M, Vernon M. A Generalized Timed Petri Net Model for Performance Analysis. IEEE Trans. Software Engineering. Vol.SE-13. No.12. Dec.1987. pp.1297-1310.

[8]   Karp RM, Miller RE. Parallel program schemata. J of Comp and Sys Sci. No.3. 1969. pp.147-195.

[9]   Mohanty S.   http://www.ece.uc.edu/~smohanty/homepage.html
University of Cininnati. 1994.

[10] Navabi Z. VHDL: Analysis and modeling of digital systems. McGraw-Hill. 1993.

[11] Perry DL. VHDL. McGraw-Hill. 1994.

[12] Peterson JL. Petri Net Theory and the Modeling of Systems. Prentice Hall. 1981.

[13] Petri Net Homepage - http://www.daimi.aau.dk/PetriNets/tools/
University of Aarhus. 1998.

[14] Protheroe D. Design automation based upon a distributed self-timed architecture. Proc. UK IT 90. Univ. of Southampton. IEE Conf. Pub. 316. 1990.

[15] Rosenstiel W. Camposano R. Synthesising circuits from behavioral level specifications. in: Computer Hardware Description Languages and their Application. Koomen, CJ. Elsvier. 1985.

[16] Salic Z, Smailagic A. Digital systems design and prototyping using field programmable logic. Kluwer. 1997.

[17] Semenov A, Koelmans AM, Lloyd L, Yakovlev A. Designing an asynchronous processor using Petri nets. IEEE Micro. Mar/Apr. 1997. pp. 54-64.

[18] Sentovich EM, et al. SIS: A system for sequential circuit synthesis. Research memorandum UCB/ERL M92/41. University of California. 1992. (Available via anonymous ftp from ic.eecs.berkeley.edu)

[19] Skahill K. VHDL for programmable logic. Addison Wesley. 1996.

[20] Sutherland IE. Micropipelines. Comm. ACM. Vol.32. No.6. 1989. pp. 720-738.

[21] Trickey H. Flamel: A high-level hardware compiler. IEEE Trans. CAD. Vol. CAD-6. No.2. Mar.1987. pp.259-268.

V
ARCHITECTURE MODELLING AND PERFORMANCE ANALYSIS

Chapter 13

PERFORMANCE ANALYSIS OF ASYNCHRONOUS CIRCUITS AND SYSTEMS USING STOCHASTIC TIMED PETRI NETS

Aiguo Xie
Cadence Design Systems, Inc.
555 River Oak Parkway, MS 2A2
San Jose, CA 94135, USA
axie@cadence.com

Peter A. Beerel
Department of Electrical Engineering – Systems
University of Southern California
Los Angeles, CA 90089, USA
pabeerel@eiger.usc.edu

Abstract   This chapter describes and extends a recently developed approach for analyzing the performance of asynchronous circuits using stochastic timed Petri nets (STPNs) with unique- and free-choice and general delay distributions. The approach uses finite net executions to derive closed-form expressions for both upper and lower bounds of the performance metrics. The expressions can be efficiently evaluated using standard statistical methods. The mean of the upper and lower bounds provides an performance estimate which has a well-defined error interval. The error interval can often be made arbitrarily small by analyzing longer net executions at the cost of additional run-time. Experiments of several asynchronous circuits demonstrate the efficiency of the approach as well as the high quality of the estimates. The experiments include a full-scale STPN model of Intel's asynchronous instruction length decoding and steering unit with over 900 transitions and 500 places.

A. Yakovlev et al.(eds.), Hardware Design and Petri Nets, 239-268.
© 2000 *Kluwer Academic Publishers.*

Keywords: Asynchronous circuits and systems, ergodicity, performance bounds and estimates, random sampling, stochastic timed Petri nets, time separations of events

1.   INTRODUCTION

In hardware, an algorithm can be implemented with asynchronous circuits or synchronous circuits or some combination of the two. Synchronous circuits use a single global clock to sequence the operations of datapath components. In contrast, asynchronous circuits sequence the datapath operations using communicating control circuits often driven by completion sensing logic that when a datapath component has completed its operation. Synchronous design techniques are by far the predominant methodology used today because they are more mature and supported by well-developed computer-aided design tools. However, asynchronous circuits can sometimes achieve significantly higher performance than synchronous counterparts. This is because 1) under favorable input data distributions, well-designed asynchronous datapath components, such as adders and multipliers, can provide better performance, 2) asynchronous micro-architecture provides finer-grain flexibility to optimize typical datapath operation sequences and 3) asynchronous circuits automatically adapt to changes in the input data rate, varying voltage supply levels, and chip temperatures (e.g.,[21]). One dramatic design in which these potential advantages have been realized is an asynchronous x86 instruction length decoder which is 3 times faster than the comparable synchronous state-of-the-art [25].

The performance of an asynchronous design depends on three factors: the delay distributions of individual datapath components, the overhead of completion sensing and control overhead, and its micro-architecture and control flow. An optimized micro-architecture (e.g., with a proper degree of pipelining) is critical to obtaining high performance, but is difficult to achieve because it must trade off control complexity with the amount of concurrency. Currently, asynchronous designers typically rely on slow simulation of detailed designs (which can be performed only late in the design cycle) or on ad-hoc back-of-the-envelope analysis of high-level architectures (e.g., [25]). This chapter describes a recently developed automated technique aimed at guiding architectural design early in the design cycle and optimizing control circuits later in the design cycle.

	Exact	Bounds
Analytical	√	√
Simulation/ Random Sampling	√	×

Figure 13.1 A classification of performance analysis techniques for timed Petri nets.

1.1 PREVIOUS WORK

Let us first review related efforts on performance analysis. To aid the exposition, we classify them into four categories depending on their solution methods as shown in Fig. 13.1.

The techniques in the first category use analytical methods to obtain exact measures of performance metrics. In [24], Ramamoorthy et al used Karp's theorem to determine the average throughput of a timed marked graph with deterministic firing delays. In similar work, Burns [5] modeled asynchronous circuits using an event-rule system similar to timed marked graphs with fixed delays and developed a method using linear programming to determine the average cycle time of events. Lee [18] extended Burns' work to handle circuit models with OR-causality. Although computationally efficient (often with a complexity ploynomial in the model size), these techniques are severely limited because they are only applicable to deterministic systems (i.e. with only fixed delays and no choice).

To handle choice, Kudva et al [17] adopted Generalized Timed Petri Nets (GTPNs) [15] to model asynchronous circuits in which component delays modeled using their mean values. The model is then converted into a continues time Markov chain and solved for its stationary probability distribution. In other Petri net based approaches, the delays are assumed to be either fixed (possibly zero) or exponentially distributed and the analytical tool is either a discrete or continuous time Markov chain (see e.g., [20]). In addition, the size of the models that can be analyzed using these techniques is limited by the state explosion problem.

To handle arbitrarily distributed delays, Xie and Beerel proposed a time-discretized model [28, 30]. The model is converted into a discrete-time Markov chain (DTMC) and solved for its stationary distribution. To mitigate the state-explosion problem, a technique called state-compression was developed [30] to speed up the computation of the stationary distribution. Specifically, the technique reduced the state space of the DTMC to one of its feedback vertex sets. Other efforts to mitigate the state explosion problem include Buchholz's work on (exact and weak) lumpability (e.g.,[4]). Nevertheless, even with these advances, the state explosion problem has limited Markov chain based approaches to small models.

To circumvent the state explosion problem, Greenstreet et al [12] have studied the performance of closed asynchronous ring structures using queuing models. Queuing models, however, are less expressive and thus are applicable only to a relatively special class of asynchronous circuits. Besides, there is no simple solution to such models with general delay distributions.

An alternative approach to avoid the state explosion problem is to use analytical methods that yield bound on performance metrics. In [6], Campos et al proposed a method to compute throughput bounds in a limited class of timed Petri nets. These bounds are defined by considering all possible delay distributions with given means. Unfortunately, these bounds are often too wide to be practically useful. Other work has focused on computing extreme values (tight bounds) of time separation between events (TSEs) in system models where delays are expressed as intervals. Such bounds on TSEs, although typically targeted at verification applications, may serve as bounds on average performance metrics. For example, Ebergen and Berks [10] studied extreme and amortized response times of asynchronous pipelines with tree structures. Hulgaard and Burns [16] give extreme TSEs for timed Petri nets with unique- and free-choice. These techniques avoid an explicit timed state analysis, and are thus computationally efficient. For more general Petri nets, however, all known techniques requires a timed state space analysis to compute extreme TSEs (e.g.,[1]).

In contrast to the above analytical techniques, several simulation based methods have been developed. These methods compute statistical estimates of exact performance metrics in models with general delay distributions. For regenerative timed Petri nets, Shelder et al [26] proposed using independent replicants whereas Glynn and Iglehart [11] proposed using standardized time series to derive performance estimates. For non-regenerative nets with arbitrary delay distributions, Haas [14] recently proposed a performance estimation method using standardized timed series. Unfortunately, we are not aware of any available tool which implements Haas' method and consequently its computational efficiency is unclear.

1.2   OVERVIEW OF OUR APPROACH

Lastly, we come to our approach in which simulation, or more accurately random sampling, is used to derive performance bounds. In particular, we use a combination of random sampling and longest path

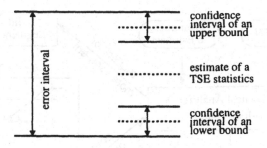

Figure 13.2 Relationship among the estimates.

analytical analysis. The major contributions of this approach are the following:

- Development of a general framework of performance metrics (e.g., throughput, latency, response time) based on a unifying notion of time separation of events (TSEs).

- Definition of TSE statistics (i.e., average, variance, and distribution of TSEs) in STPNs with unique- and free-choice and arbitrary distributed delays.

- Characterization of a large class of transition pairs for which their TSE statistics exist.

- Development of closed-form expressions for upper and lower bounds on the TSE statistics.

- Explanation of how to make above bounds sharper using more complicated expressions at the expense of additional run-time.

- Statistical estimation of the above bounds by evaluating their closed-form expressions with arbitrary high accuracy and confidence. The mean of the estimated upper and lower bounds yields an estimate of the TSE statistic with a well-defined error interval, as illustrated in Fig. 13.2.

- Demonstration of the efficiency of the approach in a variety of examples, including a performance analysis of a full-scale STPN of Intel's asynchronous instruction length decoding and steering unit, RAPPID, (with over 900 transitions) in less than 1.5 hours of CPU time.

The theoretical foundation to derive bounds on TSE statistics is based on a notion of segments of net executions. Roughly speaking, a segment is an untimed event graph corresponding to an untimed execution of

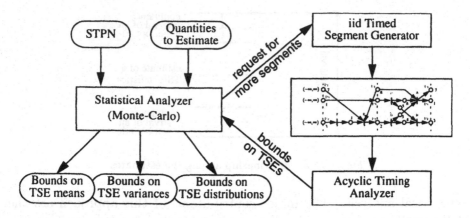

Figure 13.3   Overview of PET.

the STPN that starts and ends with the same marking. For the STPN considered in this chapter, a random time execution can be partitioned into a sequence of independent and identically distributed (iid) segments. Within each segment, a lower and an upper bound on the TSE instance is derived using a simple longest path analysis. As a result, we obtain a iid samples of the lower and upper bounds which can be statistically analyzed to obtain an unbiased estimate of the bounds. Fig. 13.3 depicts a fully automated tool named PET that implements this approach.

1.3   ORGANIZATION

The remainder of the chapter is organized as follows. Section 2 describes the class of STPNs we consider, its relationship with other STPN models commonly referred to in the literature, and the definitions of TSE statistics. Section 3 formally introduces the concept of segments and their properties, and the algorithms to compute bounds on TSEs within individual segments. Sections 4 and 5 describe the techniques to obtain bounds on various statistics of TSEs including their averages, variance and distributions. Experiments are discussed in 6. Finally, we conclude the chapter in Section 7. More formalized description of some key theorems can be found in an earlier version of this work [31].

2.   STPN MODELS OF ASYNCHRONOUS SYSTEMS

2.1   PROBABILISTIC MODELING

There have been developed many formal specifications of asynchronous systems. Traditionally, these formalism targets synthesis and

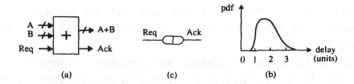

Figure 13.4 Probabilistic modeling and time discretization: (a) a datapath component, (b) an abstraction, and (c) its delay distribution.

verification, and thus are not stochastic in nature. For instance, non-fixed component delays may be specified solely in terms of their ranges, i.e., upper and lower bounds. In order to analyze their performance, these specifications are generally inadequate and often need to be probabilistically quantified.

We focus on the derivation of our performance models from an architectural description of the system. A similar derivation is also possible from more abstract behavioral or more detailed descriptions. At the architectural level, like most hardware systems, an asynchronous system contains interacting datapath components and controllers.

In asynchronous circuits, the operations of a datapath component are typically sequenced by *go* and *done* signals that emanate from and connect to asynchronous controllers. Consequently, the delay of the done signal intimately related to system performance. In principle, this delay could be modeled as a random variable which depends on the inputs to the datapath component. However, the size of such a model would likely grow exponentially with respect to the bit-width of the datapath component, making the analysis computationally intractable. In practice, one can often achieve useful performance models by abstracting the delay of a datapath component to a random variable. The probability distribution of the random variable is estimated by taking into account the input statistics of datapath component (e.g., common v.s. rare input combinations) [28, 9] obtained through high-level algorithmic simulation. This dramatically simplifies the complexity of the system model. As an example, Fig. 13.4 (c) plots the distribution of the delay of the adder shown in Fig. 13.4 (a). In some cases, such an abstraction may lead to significant absolute error in the performance estimates. One reason is that this abstraction does not account for the correlations between the delays of different data-path components. When necessary, however, these correlations can be explicitly accounted for by using extra random variables at the expense of a more complicated system model.

In addition, datapath components often generate signals that dictate operations of control circuits, such as the result of a comparator. These signals can also be modeled as random Boolean variables whose distributions can be estimated using, for example, high-level algorithmic simulation (e.g., [9]).

The issues regarding modeling asynchronous controllers are made more obtuse because there is no universal specification language for asynchronous controllers. Regardless of the specification language used, however, the model must be augmented with delays and their distributions. For example, one performance model for asynchronous controllers (implemented as finite state machines) specifies the latency of each state transition as a constant [28]. In order to incorporate the effects of the variation in chip operating temperatures and/or the manufacturing process, the relevant delays may be modeled probabilistically. Even under constant temperature and nominal process assumptions, some controllers do exhibit non-fixed delays and should be modeled probabilistically. For instance, the delay of an arbiter in metastability may be modeled as an exponential distributed random variable with proper parameters [8]. In addition, the probability distributions of input (environmental) choices can be modeled in the same manner as we model the output signals of datapath components that dictate the operations of control circuits.

The above probabilistic specifications lead to a performance model consisting of a set of random variables which are related one to another by the system logic. To specify this system logic, we use Petri nets which when annotated with delays and choice distributions are known as stochastic timed Petri nets (STPNs). Moreover, since most asynchronous system components have bounded delays, it is important to allow the specification of general delay distributions in the STPN models.

2.2   PETRI NET MODELS

Because Petri nets have been widely and successfully used to specify asynchronous systems for both synthesis and verification, they are a natural model for performance analysis [32]. They easily capture the key aspects of concurrency, synchronization, and choice that are common in asynchronous circuits. In this section, we review the subclass of Petri nets that our technique can currently analyze. More information on Petri nets may be found in the survey paper [23].

A Petri net is a triple $N = (P, T, F)$ where P is the set of places, T the set of transitions, and $F \subseteq (P \times T) \cup (T \times P)$ the flow relation. N is a *marked graph* (also known as *event graph*) if every place has at most one input and one output transition. A place p is a *choice* place if $|p\bullet| > 1$. N

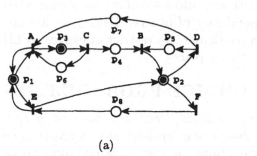

$$X(p_1) \sim \texttt{uniform}(5, 10),$$
$$X(p_2) \sim \texttt{exp}(1.5),$$
$$X(p_3) = 1,$$
$$X(p_4) \sim 1 + 5\beta(1, 3).$$

$$\cdots$$

$$\mu(p_2, D) = 0.6,$$
$$\mu(p_2, F) = 0.4 = 1 - \mu(p_2, D).$$

(a)   (b)

Figure 13.5 A stochastic Petri net example: (a) Petri net and (b) its stochastic assumption.

is a *free-choice* (FC) net if $\forall p_1, p_2 \in P$, $p_1\bullet \cap p_2\bullet \neq \emptyset \Rightarrow |p_1\bullet| = |p_2\bullet| = 1$. Equivalently, in a FC net, if two transitions share an input place, they do not have any other input places. A Petri net is *extended free-choice* (EFC) if $\forall p_1, p_2 \in P$, $p_1\bullet \cap p_2\bullet \neq \emptyset \Rightarrow p_1\bullet = p_2\bullet$. An EFC net can be translated into a FC net [23]. A choice place is asymmetric if it is neither FC nor EFC. An asymmetric choice is *unique-choice* if at most one of its output transitions can be enabled at a time. Otherwise, it is an *arbitration*.

The class of underlying (untimed) nets of the performance models we consider is the PNs with (extended) *free-choice* and/or *unique-choice*, i.e., PNs without arbitration choice. Fig. 13.5 shows an example of such a net where p_2 is free-choice, and p_1 is unique-choice since p_6, p_7 and p_8 can never have tokens simultaneously.

A marking is a mapping $M : P \rightarrow \{0, 1, 2, \cdots\}$ where $M(p)$ denotes the number of tokens in place p. A transition t is *enabled* at marking M if $M(p) \geq 1, \forall p \in \bullet t$. An enabled transition may fire. The firing of t removes one token from each place in its preset and deposits one token to each place in its poset, leading to a new marking M', denoted by $M[t\rangle M'$. A sequence of transitions $\sigma = t_0 t_1 \cdots t_{m-1}$ is a firing sequence from a marking M_0 iff $M_k[t_k\rangle M_{k+1}$ for $k = 1, \cdots, m - 1$. In this case, we write $M_0[\sigma\rangle M_m$ and say σ has a length of m denoted by $|\sigma|$. We write $\vec{\sigma}$ to be the firing counter vector of σ, indicating the numbers of times transitions are fired along σ. The vector is called a *T-invariance* if there is a marking M such that $M[\sigma\rangle M$.

A *marked* net Σ is a tuple (N, M_0), where N is a net and M_0 is its initial marking. The set of reachable markings of Σ is denoted by $R(M_0)$. Σ is *live* iff all transitions will eventually be enabled from every $M \in R(M_0)$. It is *k-bounded* if $M(p) \leq k$ for $\forall M \in R(M_0), \forall p \in P$. A 1-bounded net is also called *safe*. A live and bounded (LB) marked net has no source or sink places and no source or sink transitions (e.g., [23]). Thus, a LB net can be partitioned into a set of strongly connected components each evolving independently of others. Below, we assume the net is strongly connected. In particular, we restrict ourselves to LB nets with free-choice and unique-choice places.

2.3   TIMING AND CHOICE PROBABILITIES

For the purpose of performance analysis, Petri nets have been augmented with various types of stochastic assumptions. For example, time may be assigned to places, or transitions, or tokens, largely depending on personal taste.

To model asynchronous circuits, we found it most convenient to associate time with places. That is, a token flowing into a place p experiences a random delay associated with p before it can be consumed by an output transition of p. The actual firing of a transition is assumed to be *instantaneous*. For Petri nets with free- or unique- choices, these assumptions imply there is no race condition among transitions in structural conflict, i.e., the poset of a choice place.

For a place p, we write $X(p)$ as the random variable denoting the delay associated with it, and write $F_{X(p)} : R \to [0, 1]$ as the distribution function of $X(p)$, i.e., $F_{X(p)}(x) = \text{Prob}(X(p) \leq x)$. Unlike some other research, we do not put any restriction on the distribution functions except that they all have finite moments up to a sufficiently high order.

For each choice place p, we assume there is a probability mass function (p.m.f) $\mu(p, \cdot)$ to resolve the choice. That is, if $t \in p\bullet$, $\mu(p, t)$ is the probability that t consumes the token each time p is marked. Of course, $\sum_{t \in p\bullet} \mu(p, t) = 1$.

Example   Fig. 13.5.(b) lists a possible stochastic assumption on delays and choice p.m.f. on the net in Fig. 13.5.(a). For example, the delay on place p_1 is uniformly distributed between 5 and 10, the delay on p_2 is exponentially distributed with a parameter 1.5, the delay on p_3 is 1, the delay on p_4 is β-shaped with a parameter of $(1, 3)$ which is located at 1 and scaled by a factor of 5, and so forth. In addition, with probability 0.6, D fires each time p_2 receives a token, and with the remaining probability, F fires. The other choice place p_1 is not assigned a p.m.f. since it is unique-choice. □

One of the most general class of timed Petri nets is the GTTD_SPN where delays may be arbitrarily distributed [20]. A major difference between GTTD_SPNs and our STPN models is in the choice resolution semantics. In GTTD_SPNs, every choice, whether free-choice or asymmetric choice, is resolved through a race, i.e., the earliest firable

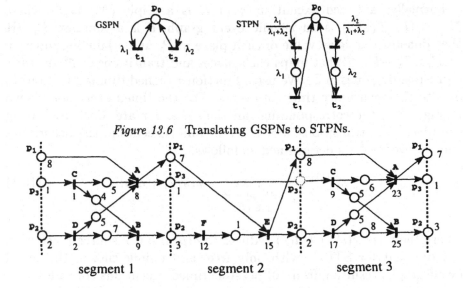

Figure 13.6   Translating GSPNs to STPNs.

Figure 13.7   A timed execution of the STPN in Fig. 13.5.

transition in the conflict consumes the tokens in the choice place. We call such a process as *dynamic* choice resolution. In contrast, our STPN models only support *static* choice resolution. Although GTTD_SPNs are very general, existing analytical solutions apply only to a subclass of them. GSPNs [19] fall into this subclass where the delays are either exponentially distributed or zero. It can be shown that a GSPN without arbitration choice can be translated into a performance equivalent STPN of ours as shown in Fig. 13.6.

2.4   TIMED EXECUTIONS

We call a possible run of an STPN a *timed execution* where choices are resolved and places are assigned delay values. In particular, we call a firing of a transition an *event*. A timed execution can be described as a sequence of events and their occurrence times. Alternatively, it can be depicted as a timed event graph which describes the causality among events.

For example, Fig. 13.7 shows the event graph of a timed execution of the STPN in Fig. 13.5. In this execution, D fires after p_2 is marked for the 1st and 3rd times whereas F fires after p_2 is marked for the 2nd time. The numbers along the (instanced) places denote the delay values assigned to them. For convenience, we write $t^{(k)}$ and $p^{(j)}$ to denote the k-th event due to the firing of t and the j-th instance of place p, respectively. In the figure, the indices are dropped for brevity.

Formally, a timed execution π of Σ is a triple (N_π, d_π, ℓ) where $N_\pi = (P_\pi, T_\pi, F_\pi)$ is an acyclic event graph, d is a function $P_\pi \rightarrow \mathbf{R}$ that denotes the delay value of each place in P_π and a labeling function $\ell : P_\pi \cup T_\pi \rightarrow P \cup T$ that maps each places and transitions of N_π to their corresponding ones in Σ. We use a function τ (called timing function) to denote the occurrence times of events. For the timed execution shown in Fig. 13.7, the corresponding functions d and τ are illustrated along the places and transitions. For a given timed execution, the occurrence time of event $t^{(k)}$ is determined as follows:

$$\tau(t^{(k)}) = \max_{\substack{(s^{(j)}, p) \in F_\pi, \\ p \in \bullet t^{(k)}}} \tau(s^{(j)}) + d(p) \qquad (13.1)$$

where the term $\tau(s^{(j)}) + d(p)$ reduces to $d(p)$ if p is a source place of π.

Note that for STPNs with only free- and unique-choice, the set of event graphs resulting from all possible timed executions coincides with the set of all possible untimed processes of the net (e.g.,[16]). This fact will be exploited later to decouple the choice resolutions from the timing behavior.

2.5   TSES AND THEIR STATISTICS

Given a timed execution π, the Time Separation of an Event pair (TSE) is the time distance between the two events. More precisely, the TSE of event pair $(s^{(k)}, t^{(k+\varepsilon)})$, denoted by, $\gamma^{(k)}(s, t, \varepsilon)$, is:

$$\gamma^{(k)}(s, t, \varepsilon) = \tau(t^{(k+\varepsilon)}) - \tau(s^{(k)}) \qquad (13.2)$$

where τ is the timing function of π, and (s, t, ε) is called the corresponding *separation triple*. When π is viewed as a random timed execution, the TSE $\gamma^{(k)}(s, t, \varepsilon)$ is a random variable, and the sequence $\{\gamma^{(k)}(s, t, \varepsilon) : k = 1, 2, \cdots\}$ is a random process.

Definition 1 *The average TSE due to separation triple (s, t, ε), denoted by $\overline{\gamma}(s, t, \varepsilon)$, is the average of the corresponding TSEs of an infinite timed execution of the STPN. That is,*

$$\overline{\gamma}(s, t, \varepsilon) = \lim_{n \to \infty} \frac{1}{n} \sum_{k=1}^{n} \gamma^{(k)}(s, t, \varepsilon). \qquad (13.3)$$

As we will show later, many system performance metrics such as average throughput and latency can be directly expressed as the average TSEs of some indicating event pairs.

Definition 2 *The variance of the TSEs due to separation triple* (s, t, ε)
is:

$$\sigma^2_{\gamma(s,t,\varepsilon)} = \lim_{n\to\infty} \frac{1}{n} \sum_{k=1}^{n} |\gamma^{(k)}(s, t, \varepsilon) - \overline{\gamma}(s, t, \varepsilon)|^2 \qquad (13.4)$$

Equivalently, $\sigma^2_{\gamma(s,t,\varepsilon)} = \lim_{n\to\infty} \frac{1}{n} \sum_{k=1}^{n} \gamma^{(k)^2}(s, t, \varepsilon) - \overline{\gamma}^2(s, t, \varepsilon).$

Definition 3 *The frequency function of the TSEs due to separation
triple* (s, t, ε) *is a mapping* $F_{\gamma(s,t,\varepsilon)} : \mathbf{R} \to [0, 1]$ *such that*

$$F_{\gamma(s,t,\varepsilon)}(x) = \lim_{n\to\infty} \frac{1}{n} \sum_{k=1}^{n} \mathbf{1}_{\gamma^{(k)}(s,t,\varepsilon) \leq x} \qquad (13.5)$$

where the indicating function $\mathbf{1}_A$ *evaluates to 1 if the subscript expression
A is true, and 0 otherwise.*

It is the above defined statistics that we wish to characterize in this
chapter. Toward this end, it is natural to first question their existence.
There are cases where an average TSE does not exist. For instance, let us
consider transitions D and F of the Petri net in Fig. 13.5. The difference
between the numbers of occurrences of D and F is infinite almost surely
as time progresses. Consequently, their average TSE diverges almost
surely for any finite occurrence offset ε provided that the net has at
least one place with a positive mean delay.

However, for a wide class of TSE pairs, their average TSEs exist [31].
Condition 1 below formally characterizes such a class with the aid of
the notion of *steady markings*. A steady marking[1] is one that can be
reached from all reachable markings.

Condition 1 *Let t and s be two transitions of Σ. If M is a steady
marking and σ is a firing sequence such that* $M[\sigma\rangle M$, *then* $\vec{\sigma}(s) = \vec{\sigma}(t)$.

The condition requires that the net fires both transitions the same
amount of times in order to traverse any cycle of steady markings. In
practice, we expect this condition to be guaranteed by the user. However,
it can also be checked using structural analysis for free-choice nets and
a reachability analysis (e.g., [29]) for nets with both free-choice and
unique-choice.

The following theorem [31] verifies that when transitions (s, t) satisfies
Condition 1, their TSE sequence generated by a random timed execution
is *weakly ergodic* and thus its average exists.

---

[1] All the steady markings of a LB Petri net with unique- and free-choices make up the unique
strongly connected component of the reachability graph of the net [3].

Theorem 4 *Let Σ be a LB Petri net that has only free-choices and unique-choices and satisfies stochastic assumptions given in Section 2.3. For any transition pair (s,t) for which Condition 1 holds, its corresponding average TSE with a fixed occurrent offset ε is a finite constant $\overline{\gamma}(s,t,\varepsilon)$ almost surely and in mean. That is,*

$$\text{Prob}(\lim_{n\to\infty} \frac{1}{n} \sum_{k=1}^{n} \gamma^{(k)}(s,t,\varepsilon) = \overline{\gamma}(s,t,\varepsilon)) = 1, \qquad (13.6)$$

$$\lim_{n\to\infty} \frac{1}{n} \sum_{k=1}^{n} \mathbf{E}\gamma^{(k)}(s,t,\varepsilon) = \overline{\gamma}(s,t,\varepsilon). \qquad (13.7)$$

In some cases, the average time distance between some special occurrences of two transitions still exists although the two transitions do not satisfy Condition 1. Below, we identify two such cases where the approach described in this chapter is still applicable.

Case 1 *The interested average time distance is between an occurrence of s to the ε-th occurrence of t following the occurrence of s.*

Such an average time distance may make sense if s occurs less frequently than t, although it does not satisfy the definition of average TSE set earlier. If one can skip the irrelevant occurrence of t, Condition 1 is met. For instance, the pair (D,F) in Fig. 13.5.(a) can be considered within this case.

Case 2 *The pair (s,t) is such that for every T-invariant $\vec{v}$, if $\vec{v}(s) > 0$, $\vec{v}(t) > 0$, and vice versa. Moreover, $\vec{v}(s)/\vec{v}(t)$ is a constant.*

In this case, one may split the occurrences of each of the two transitions into the occurrences of some new transitions, and consider the average TSE of a pair of new transitions corresponding to s and t. The new transition pair satisfies Condition 1.

Below, we assume the interested transition pair satisfies Condition 1 which serves as the basis of the other two cases.

The existence of TSE variance in the STPNs requires more consideration. In practice, the delays on places are finite or their moments are finite up to a sufficiently high order. In such cases, we believe when a TSE sequence is weakly ergodic (e.g., as characterized by Theorem 4), the existence of its variance (and in fact, any of its finite-order moments) is guaranteed. Similarly, when the TSE sequence is weakly ergodic, we believe its frequency function defined as the limit in (13.5) exists at every fixed point almost surely and in mean. In that case, we also call the frequency function as the *distribution* of the TSE.

3.   SEGMENTS AND TSE BOUNDS

The key idea of our approach is to partition an infinite random timed execution into a sequence of so called *segments* that are independent and identically distributed (iid). The targeted TSEs are then analyzed within individual segments independent of other segments. This analysis yields upper and lower bounds on the TSEs. These bounds are iid, which facilitates the estimation of the statistics of the bounds with simple statistical methods.

3.1   PARTITIONING INFINITE TIMED EXECUTIONS

Let $\pi = (N_\pi, d, \ell)$ be a timed execution of Σ. It is known that every (untimed) reachable marking of Σ induces a cut ξ (a set of instanced places) which partitions the event graph N_π [2]. The portion of the event graph in between two different cuts due to the same reachable marking M is a *segment* [31]. Formally, if σ is a firing sequence that starts from a cut ξ and ends at another cut ξ' such that $\ell(\xi) = \ell(\xi') = M$, then the portion of N_π between ξ and ξ' is a segment, denoted by $S(\xi, \sigma)$. The segment is *minimal* if it does not contain any other segment starting from ξ. For the timed execution shown in Fig. 13.7, there are three minimal segments corresponding to the reachable marking where places p_1 and p_3 are marked. Below, when we write segments, we refer to minimal ones. For convenience, we write $S(M)$ to be the set of all possible segments starting from a cut due to marking M.

One simple property of a random time execution of a Petri net considered in this chapter is that the structures of its segments are independent of each other. This is simply because the structure of a segment is determined by the choices made on the places inside the segment and choices made in different segments are independent. As a result, the sequence of segments generated by a random timed execution has a property that their structures are not determined by the location of the segment in the sequence. In fact, they are iid. This simple fact allows us to reason about an infinite execution by considering all possible finite executions of as little as one segment in length.

3.2   BOUNDING A SINGLE TSE INSTANCE

We obtain bounds on a TSE instance belonging to a particular segment by ignoring the history of the segment. In other words, our bounds on a TSE instance are defined by assuming tokens in the source places of the segment can be available at any time within $(-\infty, \infty)$. This is

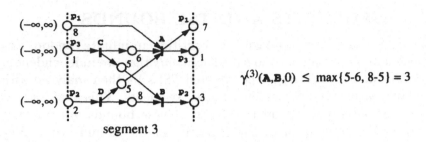

$$\gamma^{(3)}(\mathbf{A},\mathbf{B},0) \leq \max\{5\text{-}6,\ 8\text{-}5\} = 3$$

segment 3

Figure 13.8   Achieving bounds on TSE by using limited history by taking the token available time range to be $(-\infty, \infty)$ for every source place.

illustrated through the example in Fig. 13.8. This subsection describes a method to compute the upper bound using longest path analysis.

The lower bound can be computed using duality. Note that for a given ε and any $j \in N$ such that $j > -\varepsilon$, we have $-\gamma^{(j)}(s,t,\varepsilon) = -\tau(t^{(j+\varepsilon)}) + \tau(s^{(j)}) = \gamma^{(j+\varepsilon)}(t,s,-\varepsilon)$ from the definition of γ. Thus, if U is an upper bound on $\gamma(t,s,-\varepsilon)$, then $-U$ is a lower bound on $\gamma(s,t,\varepsilon)$. Because of this, we will be concerned only with the TSE upper bounds.

To compute the upper bound, we identify a set of *reference events* that serve the synchronization points for the targeted events. By assuming each of the synchronization points to be critical, one obtains a set of time separations of the event pair. The upper bound is simply the largest separation obtained. To detail this idea, consider upper bounding the TSEs due to separation triple $(s,t,\varepsilon = 0)$. Extension to the case where $\varepsilon \neq 0$ is not difficult.

Let ρ be a path in the event graph of a timed execution $\pi = (N_\pi, d, \ell)$. The set of all paths leading from x to y is denoted by $\mathcal{P}(x,y)$ where $x, y \in P_\pi \cup T_\pi$. A *reference set* for event e of π is a subset of events of π such that every path from a source place of N_π to e contains at least one event in R, and every event in R has a path to e. For example, for the segment shown in Fig. 13.7, it can be checked that $\{C^{(1)}, D^{(1)}\}$ is a reference set for $B^{(1)}$ but not for $A^{(1)}$.

It follows from the timing relation (13.1) that if x has a path to y, then y must occur after x by at least the the sum of place delays along any path from x to y. That is, whenever $\mathcal{P}(x,y) \neq \emptyset$,

$$\tau(x) + \max_{\rho \in \mathcal{P}(x,y)} \delta(\rho) \leq \tau(y). \tag{13.8}$$

where $\delta(\rho) = \sum_{p \in \rho} d(p)$. In particular, we say event x is *critical* for event y if (13.8) holds in equality. Further, if R be a reference set for event y. Then, the occurrence time of y is uniquely determined by the occurrence times of the events of R plus the delay values of places following these events. That is,

$$\tau(y) = \max_{x \in R}[\tau(x) + \max_{\rho \in \mathcal{P}(x,y)} \delta(\rho)]. \tag{13.9}$$

The term $\max_{\rho \in \mathcal{P}(x,y)} \delta(\rho)$ in (13.9) measures the maximum delay on any path from event x to y. For convenience, we write:

$$\delta^*(x,y) = \max_{\rho \in \mathcal{P}(x,y)} \delta(\rho), \tag{13.10}$$

where $\delta^*(x,y) \triangleq -\infty$ $\mathcal{P}(x,y) = \emptyset$. For completeness, we define an event e itself to be a path of delay 0, and thus $\delta^*(e,e) = 0$.

Suppose the m-th TSE instance starts in the l-th segment of π denoted by $S^{(l)}$. Since the set of source places of a segment is a cut of N_π, its poset must contain a reference set for every event e of segment $S^{(l)}$ (in fact, for every event in segments $S^{(l')}$ if $l' \geq l$). For convenience, if event $e \in S^{(l')}(l' \geq l)$, let us denote by $R(e,l)$ such a reference set. An upper bound $U_\gamma^{(m)}(s,t,0)$ on TSE $\gamma^{(m)}(s,t,0)$ is determined by (13.11) ([31]).

$$U_\gamma^{(m)}(s,t,0) = \max_{e \in R(t^{(m)},l)} [\delta^*(e,t^{(m)}) - \delta^*(e,s^{(m)})] \tag{13.11}$$

Note that the above upper bound $U_\gamma^{(m)}(s,t,0)$ is independent of the occurrence times of the events in the reference set of $s^{(m)}$. In other words, it does not depend on the history of the timed execution prior to segment $S^{(l)}$.

As an example, consider upper bounding the TSE $\gamma^{(3)}(A,B,0)$ in the segment shown in Fig. 13.8. Since $\{C^{(3)}, D^{(3)}\}$ is a reference set for $B^{(3)}$, $\gamma^{(3)}(A,B,0)$ is upper bounded by $\max\{\delta^*(C^{(3)},B^{(3)}) - \delta^*(C^{(3)},A^{(3)}), \delta^*(D^{(3)},B^{(3)}) - \delta^*(D^{(3)},A^{(3)})\} = \max\{5-6, 8-5\}=3$.

Because segments are acyclic, the TSE approximate algorithm in [7] may be used to compute the TSE upper bound. This algorithm has a time complexity no better than $O(|T(S)|^2)$ where $|T(S)|$ is the number of events in the considered segment S. However, since all the places of S have been assigned fixed delays except the source places whose delay values are unknown, the problem can be solved more efficiently. In particular, applying a longest path analysis from a fixed event e as outlined above, the term $\delta^*(e,t^{(m)})$ in (13.11) is computed in $O(|T(S)| + |P(S)|)$ time where $|P(S)|$ is the number of places in S. Thus, $U_\gamma^{(m)}(s,t,0)$ is computed in $O((|T(S)| + |P(S)|) * |R|)$ time where $|R|$ is the size of the referent set of e. This yields a typically much faster algorithm than using the one in [7]. Alternatively, the longest path analysis can be conducted from each of the targeted events back towards their reference event. This way, $U_\gamma^{(m)}(s,t,0)$ is computed in $O(|T(S)| + |P(S)|)$ time.

4.   BOUNDING TSE STATISTICS

4.1   GROUPING OF TSES

We know that the structures of segments are independent. However, multiple TSEs due to a same separation triple may start in one segment. These TSEs can be dependent. To overcome this dependency, we treat all the TSE instances starting from a same segment together as a *group*.

Consider a TSE sequence $\{\gamma^{(j)}(s,t,\varepsilon) : j \geq 1\}$ generated by a random timed execution π of $\Sigma = (N, M_0)$. The segment sequence $\{S^{(l)} \in \mathcal{S}(M_0) : l \geq 1\}$ of π partitions the TSE sequence into subsequences which we call groups. Two TSEs $\gamma^{(i)}(s,t,\varepsilon)$ and $\gamma^{(j)}(s,t,\varepsilon)$ are belong to a group if $s^{(i)}$ and $s^{(j)}$ are in a same segment. The number of TSEs of the triple (s,t,ε) in group k is called the *length* of the group, denoted by $\alpha^{(k)}(s)$. If $\eta^{(k)}(s)$ is the index of the last TSE in group k, we have

$$\alpha^{(k)}(s) = \eta^{(k+1)}(s) - \eta^{(k)}(s).$$

Let us further define the sum of all TSEs in group k to be k-th *grouped TSE*, denoted by $\beta^{(k)}(s,t,\varepsilon)$. That is

$$\beta^{(k)}(s,t,\varepsilon) = \sum_{l=\eta^{(k-1)}(s)+1}^{\eta^{(k)}(s)} \gamma^{(l)}(s,t,\varepsilon).$$

It is shown that both the group length sequence and the grouped TSE sequence are weakly ergodic [31]. In other words,

$$\frac{1}{n}\sum_{k=1}^{n} \beta^{(k)}(s,t,\varepsilon) \ \rightarrow \ \overline{\beta}(s,t,\varepsilon), \tag{13.12}$$

$$\frac{1}{n}\sum_{k=1}^{n} \alpha^{(k)}(s) \ \rightarrow \ \overline{\alpha}(s). \tag{13.13}$$

The convergence in (13.12) and (13.13) takes almost surely and in mean.

Additionally, $\overline{\beta}(s,t,\varepsilon) = \overline{\alpha}(s)\overline{\gamma}(s,t,\varepsilon)$. Note that because the structures of segments of a random timed execution are iid, the average group length $\overline{\alpha}(s)$ is merely the average length of group k (k fixed) over all possible executions, i.e., $\overline{\alpha}(s) = \mathbf{E}\alpha^{(k)}(s)$. Section 5 uses a basic statistical method to estimate $\mathbf{E}\alpha^{(k)}(s)$ as well as the bounds on $\overline{\beta}(s,t,\varepsilon)$, from which we obtain estimates of bounds on $\overline{\gamma}(s,t,\varepsilon)$.

Similarly, we define the grouped TSE-squares as the sum of squares of all the TSEs in a group. Similar to the argument of the existence of the TSE variance, we expect that the sequence of the grouped TSE-squares are also weakly ergodic.

4.2   BOUNDING THE AVERAGE TSE

Following the argument in the previous section, the problem of bounding the average TSE is translated into bounding the average grouped TSE. Note that a grouped TSE is upper bounded by the sum of the upper bounds on the TSEs in the group. For a grouped TSE $\beta^{(k)}(s, t, 0)$, we have

$$\beta^{(k)}(s, t, 0) \leq U\beta^{(k)}(s, t, 0) = \sum_{m=\eta^{(k-1)}+1}^{\eta^{(k)}} U_\gamma{}^{(m)}(s, t, 0) \qquad (13.14)$$

Thus, the expectation $\mathbf{E}U_\beta{}^{(k)}(s, t, 0)$ upper bounds the expectation $\mathbf{E}\beta^{(k)}(s, t, 0)$, both taken over all possible timed executions. Because the grouped TSE sequence is weakly ergodic, $\mathbf{E}U_\beta{}^{(0)}(s, t, 0)$ upper bounds $\overline{\beta}(s, t, 0)$.

4.3   BOUNDING MULTIPLE TSE AVERAGES

In some applications such as in performance-driven synthesis, it is desirable to find average TSEs for multiple transition pairs.

Towards this end, one approach is to apply the single-source longest path analysis from the targeted events as suggested in Section 4.2, once for each of the transition pairs. This way, for z pairs of transitions, the bounds on their TSEs in a segment S can be computed in $O((|T(S)| + |P(S)|) \times z)$ time.

For a large W, a more efficient approach is to apply the single-source longest path analysis forward from each of the poset events of source places of segment S. That way, the longest path analysis is performed $|R|$ times where R is the poset of the source places of S. Hence, the TSE bounds can be computed in $O((|T(S)| + |P(S)|) \times |R| + z)$ time.

4.4   BOUNDING TSE VARIANCE

The following inequality gives the bounds on $\sigma^2(s, t, \varepsilon)$:

$$\lim_{n \to \infty} \frac{1}{n} \sum_{k=1}^{n} \min\{U_\gamma{}^{(k)^2}(s, t, \varepsilon), L_\gamma{}^{(k)^2}(s, t, \varepsilon)\} - \max\{\overline{U_\gamma}^2(s, t, \varepsilon), \overline{L_\gamma}^2(s, t, \varepsilon)\}$$

$$\leq \sigma^2_{\gamma(s, t, \varepsilon)} \leq$$

$$\lim_{n \to \infty} \frac{1}{n} \sum_{k=1}^{n} \max\{U_\gamma{}^{(k)^2}(s, t, \varepsilon), L_\gamma{}^{(k)^2}(s, t, \varepsilon)\} - \min\{\overline{U_\gamma}^2(s, t, \varepsilon), \overline{L_\gamma}^2(s, t, \varepsilon)\}$$

where $\overline{U_\gamma}(s,t,\varepsilon)$ and $\overline{L_\gamma}(s,t,\varepsilon)$ are the means of the upper and lower bounds on the TSE, respectively. Should the above lower bound be negative, we replace it by zero.

By grouping the squares of the upper and lower bounds on individual TSEs (cf. Section 4.1), we avoid the dependency among upper and lower bounds on the TSEs of a same segment. Consequently, the grouped sum of squares of the bounds are iid due to the iid property of the random segments.

4.5   BOUNDING TSE DISTRIBUTIONS

For the k-th TSE of the triple (s,t,ε), we have

$$\mathbf{1}_{U_\gamma{}^{(k)}(s,t,\varepsilon)\leq x} \leq \mathbf{1}_{\gamma^{(k)}(s,t,\varepsilon)\leq x} \leq \mathbf{1}_{L_\gamma{}^{(k)}(s,t,\varepsilon)\leq x} \qquad (13.15)$$

where $U_\gamma{}^{(k)}(s,t,\varepsilon)$ and $L_\gamma{}^{(k)}(s,t,\varepsilon)$ are the upper and lower bounds on the TSE $\gamma^{(k)}(s,t,\varepsilon)$, respectively. The quantities $\mathbf{1}_{U_\gamma{}^{(k)}(s,t,\varepsilon)\leq x}$ and $\mathbf{1}_{L_\gamma{}^{(k)}(s,t,\varepsilon)\leq x}$ can be again determined using longest path analysis. Finally, using a technique similar to the one used in bounding the means of TSEs, we obtain bounds on $F_{\gamma(s,t,\varepsilon)}(x)$ for a fixed x. In practice, upper and lower bounds on $F_{\gamma(s,t,\varepsilon)}$ may be plotted at a sufficiently large number of points, say, 20, yielding a discretized approximation of the distribution.

5.   EVALUATING THE BOUNDS

The previous section derives closed-form expressions of the bounds on various TSE statistics. These expressions take expectations of potentially complicated random variables. In this section, we use the well-known Monte-Carlo approach (e.g.,[22]) to evaluate these expressions to a given accuracy.

5.1   MONTE-CARLO SAMPLING

Let us write W to denote the above random variable for which the expectation is taken. Let F_W be its distribution function.

The intuition behind Monte-Carlo approach is the Central Limit Theorem [13] which characterizes the partial sum of iid random variables. In our case, if $W_1, W_2, \cdots, W_n$ are iid random variables distribution F_W, the random variable $\frac{1}{n}S_n$ approaches $\mathbf{E}W$ as n grows, where $S_n = W_1 + W_2 + \cdots + W_n$. More precisely, $\frac{1}{n}S_n$ approaches the *normal* distribution with mean $\mathbf{E}W$ and variance $\frac{1}{n}\sigma_W^2$ where σ_W is the variance of W. Since $\frac{1}{n}\sigma_W^2$ decreases to 0 as n grows, any *realization* of the random variable $\frac{1}{n}S_n$ is a good (unbiased) estimate of $\mathbf{E}W$ when n is large.

In fact, for a given *relative error interval* I_e and a confidence level L_c, $P\{|(\frac{1}{n}S_n - \mathbf{E}W)/\mathbf{E}W| < I_e\} > L_c$ if

$$n > (\frac{z_{x/2}\sigma_W}{I_e\mathbf{E}W})^2, \qquad (13.16)$$

where $x = 1 - L_c$ and $z_{x/2}$ is defined such that the tail probability to its right under the standard normal distribution is $x/2$. Such a realization of $\frac{1}{n}S_n$ is known as *Monte-Carlo* sampling. It randomly realizes n iid variables (W_1 through W_n), yielding a *sample* of W of size n.

In practice, the variance of W, i.e., σ_W^2 is unknown and is commonly replaced by the variance of the sample, i.e., $S^2 = \frac{1}{n-1}\sum(W_i - \frac{1}{n}S_n)^2$. Similarly, $\mathbf{E}W$ is replaced by the mean of the sample, i.e., $\frac{1}{n}S_n$ itself. Consequently, $z_{x/2}$ in (13.16) with $x = 1 - L_c$ has to be replaced by $t_{x/2}$ which is defined such that the tail probability to its right under the t-distribution [22] is $x/2$.

Strictly speaking, the applicability of Monte-Carlo approach to our analysis also depends on the finiteness of high order moments of the delays of the STPNs. However, we do not believe this is a significant restriction because in real designs, delays of system components are either finite or their moments are finite up to a sufficiently high order.

5.2   IMPROVING THE BOUNDS

In many cases, the upper and lower bounds on the TSE statistics derived in the previous subsections are very close to each other, which combined yield very good estimates. Nevertheless, the lower and upper bounds can be made even closer to each other by using some amount of *history* segments prior to the segments being considered. This effectively pushes the reference set further apart from the target TSE pairs. Fig. 13.9 demonstrate this improvement compared with the bound achieved without any history segment (see Fig. 13.8).

6.   EXPERIMENTS

The bounding technique described in this chapter has been fully automated, resulting a tool named PET. In this section, we apply PET to analyze two scalable examples. The first example is a STPN model of the micropipeline [27] where each stage exhibits choice, and the second a STPN model of Intel's asynchronous instruction length decoding and steering unit RAPPID [25].

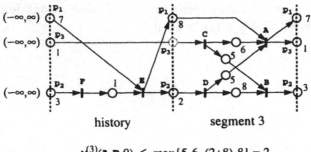

$$\gamma^{(3)}(\mathbf{A},\mathbf{B},0) \leq \max\{5\text{-}6, (2\text{+}8)\text{-}8\} = 2$$

Figure 13.9 Improving TSE bounds by adding extra segments as history.

6.1   MICROPIPELINES WITH CHOICE

Fig. 13.10 depicts a STPN model of a micropipeline. At each stage i, the delays of places p_{i_1} and p_{i_4} are uniformly distributed between 0 and 10, and the delays of places p_{i_2} and p_{i_6} are uniformly distributed between 1 and 2. Places p_{i_3} and p_{i_5} assume a fixed delay of 2. The other two places p_{i_7} and p_{i_8} assume a fixed delay of 0.5. In addition, each choice place has two output transitions and has a choice p.m.f. that evaluates to 0.5 at each of the output transition.

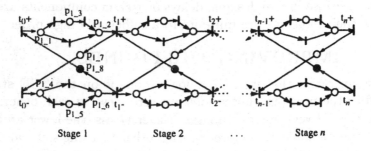

Figure 13.10 A micropipeline where each stage exhibits choice.

Table 13.1 lists the estimated bounds on the mean and variance of the cycle time of the pipeline (i.e., $\gamma(t_i+, t_i+, 1)$ with i fixed) as a function of the pipeline depth (i.e., the number of pipeline stages). It is interesting to note that the sample size needed in estimating the bounds of variance of the cycle time is remarkably larger ($3 \sim 5$ times larger) than that needed in estimating mean of the cycle time, and thus more CPU time is required. In the table, the amount of history segments is set to approximately 1.5 times the pipeline depth. The results also show that the cycle time of the pipeline increases slightly in mean but decreases dramatically in variance as the pipeline depth increases.

Table 13.1 Bounds on the mean and variance of cycle time of the micropipeline. The relative error interval and confidence level are set to 1% and 99%, respectively.

# stages	# transitions	# places	Mean			Variance		
			Lower bound	Upper bound	CPU time (secs)	Lower bound	Upper bound	CPU time (secs)
2	18	16	19.35	19.35	2.8	12.69	12.69	3.9
4	32	32	21.78	21.78	4.3	7.05	7.21	9.2
8	66	64	22.67	22.68	8.1	5.76	5.90	43.5
16	130	128	23.11	23.12	34.0	5.41	5.58	113.2
32	258	256	23.55	23.55	148.8	4.58	5.31	781.8

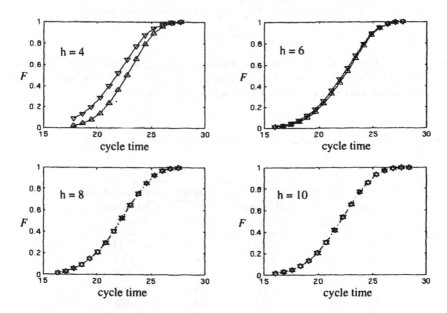

Figure 13.11 Improvement of bounds on the cycle time distribution by increasing the number of history segments (h). The relative error interval and confidence level are set to 1% and 99%, respectively.

Fig. 13.11 shows 4 plots of the lower and upper bounds on the cycle time distribution of an 8-stage micropipeline as the amount of added history segments increases. As in the case of bounding the mean and variance, increasing the amount of considered history dramatically improves the the bounds on the distribution function. Fig. 13.12 plots the estimated distribution function of the cycle time of pipelines with different depths. As the depth increases, the distribution function shifts towards the right.

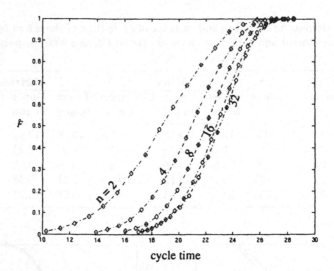

Figure 13.12 The estimated cycle time distributions (the average of the upper and lower bounds) of pipelines with different depths (n is the number of stages). The relative error interval and confidence level are set to 1% and 99%.

Table 13.2 Computing bounds on multiple average TSEs: bounding average latencies.

# stages	through entire pipeline		through individual stages	
	# TSEs	CPU time (secs)	# TSEs	CPU time (secs)
2	1	3.29	2	3.44
4	1	9.22	4	11.6
8	1	28.5	8	32.8
16	1	111	16	119
32	1	446	32	460

The efficiency of the tool in computing bounds on statistics of multiple TSEs is demonstrated by simultaneously estimating bounds on average latencies within individual pipeline stages. Table 13.2 lists the CPU times required for both single TSE estimation (estimating the average latency through the entire micropipeline) versus those required to compute bounds on average latencies within individual stages. The CPU times are reported by applying the longest path analysis from the poset set of the source places of segments. Note that the extra percentage of CPU time required to compute bounds on multiple average TSEs is negligible. In addition, this percentage is almost in proportion to the number of separation triples being analyzed.

6.2   A MODEL OF RAPPID

The RAPPID [25] is an aggressive asynchronous design to decode the length of x86 instructions and steer them to the subsequent instruction decoder with a high average throughput. The instructions are fed into the design through a number of parallel byte units associated with columns. Each column speculatively decodes the byte under the assumption that an instruction starts at its associated byte. It has a number of tag units responsible for receiving a token indicating this column is the start of the next instruction and routing the token to a tag unit associated with the column which contains the start of the subsequent instruction. The tag units are arranged in rows each associated with an output buffer for the decoded instructions. To alleviate a bottleneck in the output buffers, tag units in one row send its token to tag units in the next row. Both columns and rows are designed in a ring fashion so that they handle an infinite number of instructions. The real design has 16 columns, 4 rows, and is optimized to decode instructions of up to 7 bytes in length.

In the experiment, we examine a live and safe Petri net model of the RAPPID where the delays of places are only loosely related to the real delays in the design. The model ignores the handling of very rare instructions (e.g., length changing prefixes) and branches. The central driving units for the model is the actual length decoding logic in the columns. These units are optimized for the common instruction lengths and consequently have longer delay for less common instructions [9]. In the model, depicted in Fig. 13.13, this logic is modeled with a free-choice place whose choice p.m.f. matches the relative frequency of instruction lengths given in [9]. Even for a given instruction length the decoding time can vary, thereby motivating the use of stochastic delay models for the decoding of each length.

Once decoded, the column broadcasts a signal to multiple tag units associated with it, indicating an instruction start at the column. When all the bytes of the instruction are ready, the instruction is routed to a proper output buffer. In our model, the tag unit in the top row of the first column is initially marked with a token to indicate that the first instruction will start at the first column and should be routed to the first output buffer. Once all bytes of the first instruction are available, this tag unit sends a token to a downstream tag unit in the second row of the column where the next instruction starts. We use unique-choice places

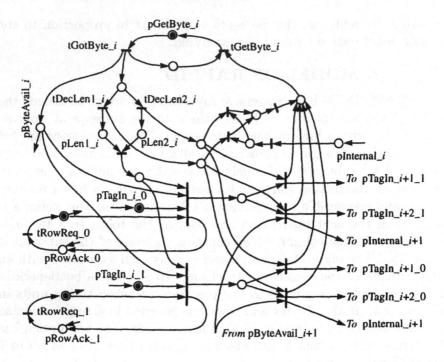

Figure 13.13 A Petri net model of 1 column of the RAPPID. The column has 2 Tag Unit rows and can handle instructions of length up to 2 bytes.

in the model to route tokens to downstream tag units as well as the consumption of tokens representing all available bytes of an instruction.

We specified uniform delays of between 0.1 and 0.6 time units for pGotBuff, between 0.5 and 1.5 time units for pLen1, pLen2, and pLen3, between 1.0 and 2.5 time units for pLen4 and pLen5, and between 2 and 4 time units for pLen6 and pLen7. The pRowReq was given a fixed delay of 0.5 time units and all other places where given fixed delays of 0.1 time units.

We experimented with a number of different RAPPID architectures by varying the number of columns, rows, and the maximum instruction lengths handled. For each architecture, the mean and variance of the time separation between consecutive firings of transition **tRowReq0**, i.e., $\overline{\gamma}(\mathtt{tRowReq0}, \mathtt{tRowReq0}, 1)$ were analyzed. The bounds on the mean of this TSE divided by the number of rows provides an estimate of the average decoding cycle time and thus the throughput. The bounds on its variance give an estimate of the performance deviation. Table 13.3 shows the experimental results. The listed CPU times were required to bound the variance, which includes the times for bounding the averages.

Table 13.3 Bounds on the mean and variance of the decoding cycle time of the RAPPID model. The relative error interval and confidence level are set to 5% and 99%, respectively.

# Col.	# Row	Len.	# Trans.	# Place	Mean Lower	Mean Upper	Variance Lower	Variance Upper	CPU time secs
2	1	2	29	34	1.315	1.327	0.219	0.249	40.6
2	2	2	36	40	2.643	2.644	0.437	0.442	46.9
2	3	2	43	46	3.987	3.991	0.592	0.594	55.16
2	4	2	50	52	5.295	5.304	0.753	0.754	66.89
4	1	2	57	66	0.684	0.703	0.157	0.214	156.2
4	2	2	70	76	1.374	1.376	0.222	0.231	254.3
4	3	2	83	86	2.062	2.063	0.334	0.337	153.6
4	4	2	96	96	2.744	2.746	0.361	0.363	359.5
8	1	7	273	210	0.957	0.976	0.778	0.871	411.1
8	2	7	338	228	1.925	1.926	1.420	1.428	630.5
8	3	7	403	246	2.893	2.893	2.106	2.106	990.5
8	4	7	468	264	3.843	3.843	2.689	2.690	1,385
16	1	7	545	418	0.548	0.554	0.328	0.348	558.9
16	2	7	674	452	1.056	1.056	0.821	0.824	2,189
16	3	7	803	486	1.588	1.589	1.170	1.175	3,419
16	4	7	932	520	2.122	2.123	1.355	1.358	5,426

The amount of history segments used in the experiments was set between 8 to 40 depending on the sizes of the RAPPID models.

Since the delay models do not match the real design, we would rather not stress the significance of the performance trends. However, the experiment demonstrates the low run-time of the tool and the accuracy of the estimates. For all the models analyzed, the width of the derived error bounds are less than 5% and are typically less than 1%. The largest model contains over 900 transitions and 500 places and yet can be analyzed with 5% error tolerance and 99% confidence in 1.5 hours. These experiments suggest that the tool may be used to trade off the silicon area of the design with the system throughput.

7.   CONCLUSIONS

This chapter describes some recent developments on the performance analysis of asynchronous circuits and systems using stochastic timed Petri nets. A novel technique based on random sampling approach is given to obtain precise estimates of the mean, variance and distribution of time separations of events for a wide class of transition pairs. The estimates are achieved by computing the lower and upper bounds on the targeted quantities based on iid segment generation and simple longest path analysis. Experiments on large-scale STPN models demonstrated the efficiency and capability of the tool.

The integration of the technique into performance-driven synthesis tools is under examination. The problem of extending the technique to handle STPNs with arbitration is also being studied.

Acknowledgments

This work is partially funded by NSF Grant CCR-9812164. We would like to thank the anonymous reviewers of an earlier version of this presentation for their insightful feedbacks.

References

[1] Alur, R. and Dill, D. L. (1994). A theory of timed automata. *Theoretical Computer Science*, 126:183–235.

[2] Best, E. (1990). Partial order behavior and structure of Petri nets. *Formal Aspects of Computing*, 2:123–138.

[3] Best, E. and Voss, K. (1984). Free choice systems have home states. *Acta Informatica*, 21:89–100.

[4] Buchholz, P. (1994). Exact and ordinary lumpability in finite Markov chains. *Journal of Applied Probability*, 31:59–75.

[5] Burns, S. M. (1991). *Performance Analysis and Optimization of Asynchronous Circuits*. PhD thesis, California Institute of Technology.

[6] Campos, J., Chiola, G., and Silva, M. (1991). Properties and performance bounds for closed free choice synchronized monoclass queueing networks. *IEEE Transactions on Automatic Control*, 36(12):1368–1382.

[7] Chakraborty, S. and Dill, D. L. (1997). Approximate algorithms for time separation of events. In *Proc. International Conf. Computer-Aided Design (ICCAD)*, pages 190–194.

[8] Chaney, T. J. and Molnar, C. E. (1973). Anomalous behavior of synchronizer and arbiter circuits. *IEEE Transactions on Computers*, C-22(4):421–422.

[9] Chou, W.-C., Beerel, P. A., Ginsoar, R., Kol, R., Myers, C. J., Rotem, S., Stevens, K., and Yun, K. Y. (1998). Average-case optimized technology mapping of one-hot domino circuits. In *Proc.*

International Symposium on Advanced Research in Asynchronous Circuits and Systems (ASYNC), pages 80–91. IEEE Computer Society Press.

[10] Ebergen, J. and berks, R. (1999). Reponse-time of asynchronous linear pipelines. *Proceedings of the IEEE*, 87(02):308–318.

[11] Glynn, P. W. and Iglehart, D. L. (1990). Simulation output analysis using standardized time series. *Math. Oper. Res.*, 3:1–16.

[12] Greenstreet, M. R. and Steiglitz, K. (1990). Bubbles can make self-timed pipelines fast. *Journal of VLSI Signal Processing*, 2(3):139–148.

[13] Grimmett, G. R. and Stirzaker, D. R. (1992). *Probability and Random Processes (2nd Edition)*. Oxford Science Publications.

[14] Haas, P. J. (1997). Estimation methods for stochastic Petri nets based on standardized time series. In *International Workshops on Petri Nets and Performance Modelds*, pages 194–204.

[15] Holliday, M. A. and Vernon, M. Y. (1987). A generalized timed Petri net model for performance analysis. *IEEE Transactions on Software Engineering*, 13(12):1297–1310.

[16] Hulgaard, H. and Burns, S. M. (1994). Bounded delay timing analysis of a class of CSP programs with choice. In *Proc. International Symposium on Advanced Research in Asynchronous Circuits and Systems (ASYNC)*, pages 2–11. IEEE Computer Society Press.

[17] Kudva, P., Gopalakrishnan, G., Brunvand, E., and Akella, V. (1994). Performance analysis and optimization of asynchronous circuits. In *Proc. International Conf. Computer Design (ICCD)*, pages 221–225.

[18] Lee, T. (1995). *A General Approach to Performance Analysis and Optimization of Asynchronous Circuits*. PhD thesis, California Institute of Technology.

[19] Marsan, M. A., Balbo, G., Conte, G., Donatelli, S., and Franceschinis, G. (1995). *Modelling with Generalized Stochastic Petri Nets*. John Wiley & Sons.

[20] Marsan, M. A., Bobbio, A., and Donatelli, S. (1998). Petri nets in performance analaysis: An introduction. In *Lectures on Petri nets I: Basic Models*, Lecture Notes in Computer Science. Springer-Verlag.

[21] Martin, A. J., Burns, S. M., Lee, T. K., Borkovic, D., and Hazewindus, P. J. (1989). The design of an asynchronous microprocessor. In Seitz, C. L., editor, *Advanced Research in VLSI: Proceedings of the Decennial Caltech Conference on VLSI*, pages 351–373. MIT Press.

[22] Miller, I. R., Freund, J. E., and Johnson, R. (1990). *Probability and Statistics for Engineers*. Prentice Hall.

[23] Murata, T. (1989). Petri nets: Properties, analysis and applications. *Proceedings of the IEEE*, 77:541–580.

[24] Ramamoorthy, C. V. and Ho, G. S. (1980). Performance evaluation of asynchronous concurrent systems using Petri nets. *IEEE Transactions on Software Engineering*, 6(5):440–449.

[25] Rotem, S., Stevens, K., Agapiev, B., Dike, C., Roncken, M., Ginosor, R., Kol, R., Beerel, P. A., Yun, K. Y., and Myers, C. J. (1999). Rappid: An asynchronous instruction length decoding and steering unit. In *Proc. International Symposium on Advanced Research in Asynchronous Circuits and Systems (ASYNC)*.

[26] Shedler, G. S. (1993). *Regenerative Stochastic Simulation*. Academic Press, New York.

[27] Sutherland, I. E. (1989). Micropipelines. *Communications of the ACM*, 32(6):720–738.

[28] Xie, A. and Beerel, P. A. (1997). Symbolic techniques for performance analysis of asynchronous systems based on average time separation of events. In *Proc. International Symposium on Advanced Research in Asynchronous Circuits and Systems (ASYNC)*, pages 64–75. IEEE Computer Society Press.

[29] Xie, A. and Beerel, P. A. (1998). Efficient state classification of finite state Markov chains. *IEEE Transactions on Computer-Aided Design*, 17(12):1334–1338.

[30] Xie, A. and Beerel, P. A. (1999). Accelerating Markovian analysis of asynchronous systems using state compression. *IEEE Transactions on Computer-Aided Design*, 18(7):869–888.

[31] Xie, A., Kim, S., and Beerel, P. A. (1999). Bounding average time separation of events in stochastic timed Petri nets with choice. In *Proc. International Symposium on Advanced Research in Asynchronous Circuits and Systems (ASYNC)*.

[32] Yakovlev, A. and Koelmans, A. M. (1998). Petri nets and digital hardware design. In *Lectures on Petri nets II: Basic Models*, Lecture Notes in Computer Science. Springer-Verlag.

Chapter 14

PERFORMANCE ANALYSIS OF DATAFLOW ARCHITECTURES USING TIMED COLOURED PETRI NETS

B.R.T.M. Witlox

Eindhoven University of Technology, P.O. Box 513, 5600 MB Eindhoven, The Netherlands

Philips Research Laboratories, Prof. Holstlaan 4, 5656 AA Eindhoven, The Netherlands

witloxb@natlab.research.philips.com

P. van der Wolf

Philips Research Laboratories, The Netherlands

E.H.L. Aarts

Eindhoven University of Technology, The Netherlands

Philips Research Laboratories, The Netherlands

W.M.P. van der Aalst

Eindhoven University of Technology, The Netherlands

Abstract   We present an approach to model *dataflow architectures* at a high level of abstraction using *timed coloured Petri nets*. We specifically examine the value of Petri nets for evaluating the *performance* of such architectures. For this purpose we assess the value of Petri nets both as a modelling technique for dataflow architectures and as an analysis tool that yields valuable performance data for such architectures through the execution of Petri net models. Because our aim is to use the models for performance analysis, we focus on representing the timing and communication behaviour of the architecture rather than the functionality. A *modular* approach is used to model architectures. We identify five basic hardware *building blocks* from which Petri net models of dataflow architectures can be constructed. In defining the building blocks we will

269

A. Yakovlev et al.(eds.), Hardware Design and Petri Nets, 269-289.
© 2000 *Kluwer Academic Publishers.*

identify strengths and weaknesses of Petri nets for modelling dataflow architectures. A technique called *folding* is applied to build generic models of dataflow architectures. A timed coloured Petri net model of the Prophid dataflow architecture, which is being developed at Philips Research Laboratories, is presented. This model has been designed in the tool ExSpect. The performance of the Prophid architecture has been analysed by simulation with this model.

Keywords: dataflow architectures, hardware modelling, performance analysis, Petri nets

1.   INTRODUCTION

Digital video signals for multi-media applications and digital television are high-throughput, stream-based video signals. Real-time processing of digital video signals requires high performance computation. A growing number of video applications have functions for which the computation time depends highly on the applied input data. Examples of this are encoding and decoding of images in the MPEG format. So, dynamic behaviour with variable data rates and data dependent computation is becoming a characteristic of real-time digital video signal processing. Because of the processing power required for real-time digital video signal processing, video algorithms are typically executed on special ICs, called *video signal processors.*

Due to the dynamic behaviour of video signal processing algorithms we need a computational model for real-time video signal processing that can handle variable data rates and data dependent computation. The dynamic dataflow model [2] provides a formalism to describe algorithms with this kind of dynamic and parallel behaviour. A dataflow algorithm is a directed graph where the nodes represent tasks and the arcs define the data dependencies between the tasks. Data produced and consumed by the nodes is carried in tokens which flow along the arcs.

The video signal processing algorithms have to be mapped onto architectures that will perform the computations. *Dataflow architectures* [12] are particularly suited to implement this kind of stream oriented video algorithms. Architectures and methodologies for mapping video signal processing algorithms onto architectures are being developed at Philips Research Laboratories. A video signal processing architecture called the Prophid architecture [6] is being developed, which is the subject of our study.

2. PROBLEM STATEMENT

In a design process there are usually several possibilities for implementation. To be able to make the right choices, these possible implementations need to be quantified taking into account such measures as performance, chip area and power dissipation. In this paper we focus on methods for performance evaluation. These methods must permit the designer to explore the design space in search for the implementation that offers satisfactory performance.

In the application domain of dataflow architectures for video signal processing, the performance depends on the following three elements:

- the architecture,

- the video signal processing algorithms that are used, and

- the mapping of the algorithms onto the architecture.

Note that we focus on *programmable* architectures that satisfy the performance needs of a class of algorithms.

To analyse the performance of an architecture, a workload must be imposed on the architecture. A workload consists of a video processing algorithm, a mapping for this algorithm onto the architecture and streams of video samples as input data for the algorithm. With models of the architecture, the algorithms and the mapping we can analyse the performance. This is depicted schematically in Figure 14.1, which is called the *Y-chart* because of its shape.

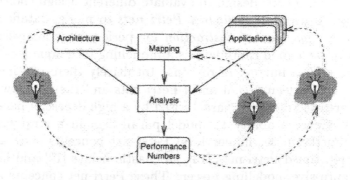

Figure 14.1 The Y-chart method for performance analysis.[5]

For the performance analysis we have to select the performance criteria to evaluate, the so-called *metrics*, that will give us a measure for the performance of the modelled architecture with an algorithm mapped onto it. Some of the metrics that are of interest to us are the utilisation of the video signal processing elements, memory usage and throughput of the system. The performance numbers that are the result of the analysis quantify the performance of the architecture and give the designer better insight in the working of dataflow architectures and the mapping of algorithms onto these architectures. This information can help the designer to improve the architecture, the way algorithms are mapped onto the architecture and the way algorithms are structured. This is represented by the feedback loops with the light bulbs in Figure 14.1. The designer repeats the analysis until he finds a satisfactory architecture design, mapping and structure for the algorithms.

Important requirements for the models and modelling techniques are:

(a) executability,

(b) ease of modelling,

(c) configurability,

(d) efficiency, and

(e) accuracy.

Requirements (a), (b) and (c) relate to the modelling technique. Requirements (d) and (e) relate to simulations with the models.

Currently design tools for performance analysis of high performance, heterogeneous systems such as dataflow architectures for video signal processing are lacking. Most available tools assist in modelling the architecture components at a detailed level. A high abstraction level is needed in an early stage of the design to evaluate different design choices.

We propose to use *timed coloured Petri nets* to model dataflow architectures at a *high level of abstraction* for performance analysis. Petri nets seem to be a good candidate as a modelling technique for dataflow architectures for a number of reasons. Intuitively there is a match between the data driven execution of Petri nets and the dataflow model. Dataflow architectures are characterised by a high degree of parallelism. Petri nets offer the ability to model parallelism in a straightforward manner. Furthermore, higher level Petri net concepts such as token colouring [4], timed Petri nets [13], stochastic delays [10] and hierarchy provide expressive modelling power. These Petri net concepts are supported by the tool ExSpect [1]. In ExSpect we can design and execute Petri nets. This enables us to run simulations with our architecture model. Computers have become much faster over the last decade which makes computer-based simulation of complex systems feasible.

We can formulate our problem statement as follows: how can timed coloured Petri nets be used to execute the Y-chart (Figure 14.1). More specifically, we want to assess the value of Petri nets for modeling dataflow architectures, video algorithms and mappings. We want to use these models to exercise the loops of the Y-chart in order to analyse the performance of different solution points in the design space. Therefore we are interested how we can configure a Petri net model with an architecture instance, an algorithm and a mapping and measure performance metrics using the Petri net model.

Several studies on performance modelling of multiprocessor systems using Petri nets have been presented in the literature [3, 8, 9, 7]. As an example we mention the work presented by [9] on performance models of multiprocessor systems. Our approach for modelling architectures for performance analysis is similar to the approach presented by Marsan. The models differ however from our model in the granularity of the tasks executed on the architectures. In most previously published work the architectures that are considered consist of processors and memory components that communicate via an interconnection structure. The performance of the architectures is analysed by modelling the exchange of messages between the components and the execution of tasks on the processors. The architectures that we focus on are stream based architectures with sample based communication between the co-processors. Therefore our model of the communication and processing of video samples is more fine-grained. To our knowledge no work has previously been published on performance modelling of dataflow architectures using Petri nets.

3.   SOLUTION APPROACH

We will assess the value of timed coloured Petri nets for modeling dataflow architectures using the requirements for models and modeling techniques that we presented in the previous section.

We model the dataflow architectures at a high level of abstraction, such that we do not have to model all hardware details and we are still able to evaluate the performance of the architectures. To model an architecture we have to model its components and the interconnections among the components. Because we are interested in the performance of the architectures, we focus on the timing behaviour of the components. We model the behaviour of the components in terms of delays and synchronisations with the activities of other components.

We use a modular approach by constructing Petri net models of basic building blocks of dataflow architectures. We use these building blocks to identify the strengths and weaknesses of Petri nets as a modelling tool for performance evaluation.

Our aim is to construct a model with a high level of genericity that can be used to simulate different instances from a class of architectures. It must be possible to configure the model with the information of a video algorithm and a mapping such that we can execute the algorithm on our model of the architecture and analyse its performance by measuring a set of metrics.

4.   ARCHITECTURE

The architecture that is subject of our study is the Prophid dataflow architecture as proposed by [6] that is being developed at the Philips Research Laboratories. A simplified structure of an architecture belonging to the class of Prophid architectures is shown in Figure 14.2.

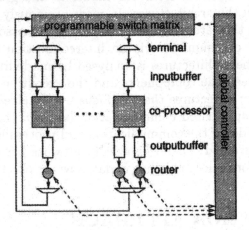

Figure 14.2   Prophid architecture.

An architecture of this class consists of a number of co-processors that autonomously process streams of video samples. The video processing algorithms that will be executed on this architecture are dynamic dataflow graphs that consist of coarse grained tasks, such as noise reduction and sample rate conversion. Tasks in the dataflow graphs are mapped onto the co-processors of the architecture. Multiple tasks may be mapped onto the same co-processor. Tasks that are mapped onto the same co-processor are executed multi-plexed in time. In dynamic dataflow, tasks can consume tokens in dynamic, variable patterns from the input edges. The number of tokens that are produced on the outputs may also vary dynamically. This results in dynamic behaviour in the co-processors for consuming samples from the input FIFO buffers and producing samples to the output FIFO buffers. The co-processors can have a computational pipeline and can have dynamic, data-dependent computation times.
The co-processors communicate via input and output FIFO buffers con-

nected with a communication structure. The communication structure consists of routers, a global controller and a switch matrix. The video sample streams are organised in sample packets. The head of a packet contains routing information and the rest of a packet contains the actual video data (samples). The routers communicate sample packets via the switch matrix from their output buffer to an input buffer of the co-processor that will perform the next task in the dataflow graph. The routers interact with the global controller to obtain the destination information.

5. HARDWARE MODELLING WITH PETRI NETS

The architectures will be realised in synchronous hardware. In hardware operations can be executed in parallel. Parallel operations can be modelled by Petri net transitions that can fire in parallel or by transitions that can fire instantaneously in a sequence. The performance numbers that will be produced by our models must be independent of possible non-determinism in the execution or evaluation of the model. Therefore we have to make sure that different possibilities for firing a sequence of transitions that are enabled simultaneously will produce the same marking and result in the same performance numbers.

We use the timed coloured Petri net model with time stamps attached to tokens and delays associated with token production [4, 11]. This form of delay is a natural choice to model dataflow architectures where the timing behaviour is determined by the delays that are involved with communication and processing of video samples. We assume in our Petri net model that the order in which tokens are placed in the places is preserved. This gives the places a FIFO property which can be exploited to model memory elements with a FIFO property.

We have identified a set of basic hardware constructs that can be seen as building blocks for the dataflow architectures of interest to us. These building blocks are:

(a) communication between 2 components (via a register),

(b) communication via a FIFO buffer,

(c) pipelined computation,

(d) merging multiple streams of samples (multiple producers, one consumer), and

(e) splitting one stream of samples into multiple streams (one producer, multiple consumers).

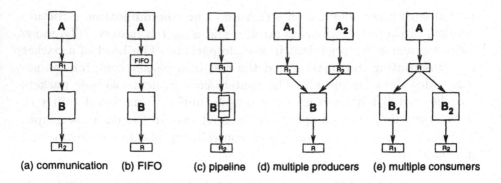

(a) communication   (b) FIFO   (c) pipeline   (d) multiple producers   (e) multiple consumers

Figure 14.3   Five basic hardware constructs.

We discuss each construct in the following subsections and show how Petri nets can be used to model the hardware constructs. Each of the sections starts with a description of the hardware construct, a description of the model, followed by a discussion of the strengths and weaknesses of Petri nets for modelling that particular hardware construct at an abstract level. With the Petri net models for the basic hardware constructs we can assemble architecture models in a modular manner by either linking building block models or by combining several building blocks to create more complex constructs.

5.1   COMMUNICATION

Hardware description.   This basic construct, shown in Figure 14.4(a), consists of two components that communicate via a register. Components A and B communicate via register R_1 and component B writes to register R_2. Write accesses to the registers have a delay of one clock cycle.

Model description.   We can model this construct with the timed coloured Petri net shown in Figure 14.4(b). The components A and B are modelled with Petri net transitions. The registers R_1 and R_2 are modelled with Petri net places. The values of the registers are modelled with "coloured" tokens that have integer values. We need time in our Petri net to model the delays in the hardware for the read and write actions from and to the registers. To ensure that transition A and B cannot fire at any given time, but only once per clock cycle, we use the selfloops containing the places *value* and *idle* to enable or disable these transitions. The delays given to the tokens in these places represent the time that a component is busy with its operation.

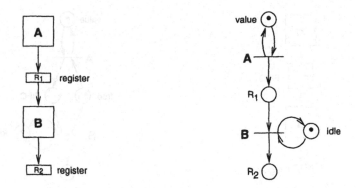

Figure 14.4   (a) Communication between components and (b) Petri net model.

Discussion.   We can conclude that concurrent read and write operations in hardware can be modelled with a timed Petri net where tokens have a time stamp. The order in which transitions fire is determined by the availability of tokens. We can use the tokens that model the values that are communicated between the hardware components, to impose an order on the firing of transitions.

Delays for communication are modelled by assigning delays to the tokens that model the communicated data. Delays for computation times are modelled by assigning delays to the tokens that represent whether a component is busy or idle.

5.2   FIFO BUFFER

Hardware description.   The FIFO buffer construct, shown in Figure 14.5(a), consists of a producer component A and a consumer component B that communicate via a FIFO buffer. Each clock cycle component A tries to write a value into the buffer. It blocks if the buffer is full. Each clock cycle component B tries to consume a value from the buffer and writes the value to register R. It blocks if the buffer is empty. Accesses to the buffer and register have a delay of one clock cycle.

Model description.   We can model the FIFO buffer with two Petri net places, one to represent the content of the buffer and one to represent the free buffer space. The Petri net model is shown in Figure 14.5(b). The delay of one clock cycle in the hardware when a value is written to the buffer is modelled by giving the token that transition A places in place *FIFO* a delay of one time unit. This also applies for the tokens that B places in *free* and R. The tokens in the selfloops containing the places *idle* and *value* determine the rate at which tokens can be consumed from the buffer and placed in the buffer.

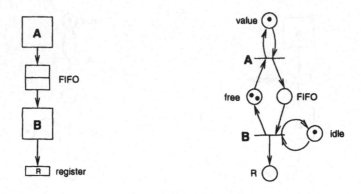

Figure 14.5   (a) FIFO buffer and (b) Petri net model.

Discussion.   Because the order of the tokens in place *FIFO* with respect to their arrival time is preserved in the timed Petri net model with time stamps, we can model the FIFO buffer in such a compact way.

5.3   PIPELINE

Hardware description.   This construct, shown in Figure 14.6(a), contains a component *B* that performs a pipelined computation. The output that it produces depends on a number of consecutive input values produced by component *A*. The pipeline is blocking for consumption of input data if it is full and it is blocking for production of output data if it is not full. Each clock cycle components *B* tries to perform its operation.

Model description.   We model the pipeline with a Petri net place *pipeline*, similar to our model of the FIFO buffer (section 5.2). The tokens that represent the data accumulate in this place. Their order is preserved by the FIFO property of the Petri net place.

We model component *B* with two transitions, *B-consume* and *B-produce* for adding and removing tokens from the pipeline respectively. If the pipeline is not full, *B-consume* consumes a token from the input place and places it in the pipeline place. The free spaces in the pipeline are modelled by tokens in a place *free*, just like we modelled the FIFO buffer. If the pipeline is full, *B-produce* removes a token from the pipeline and produces an output token. However we cannot use the *pipeline* place alone to enable transition *B-produce* when the pipeline is full, because *B-produce* does not know how many tokens reside in the *pipeline* place. To model the "block on not full" property of *B*, we need an extra place, *N*, with a token that indicates the number of tokens in *pipeline*. To

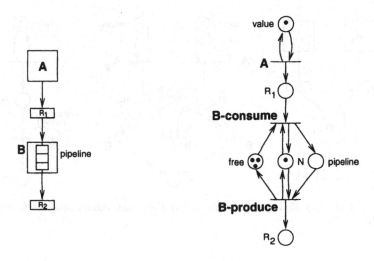

Figure 14.6   (a) Pipelined computation and (b) Petri net model.

enable *B-produce* only when the pipeline is full, we give *B-produce* the precondition $N = pipeline\text{-}size$.

Discussion.   We can model a pipeline in a similar manner as a FIFO buffer. The FIFO property of places in the timed Petri net model enables us to use a place to model the content of the pipeline. The extra place is needed to provide the transition that removes tokens from the pipeline with information about the number tokens in the pipeline.

5.4   MULTIPLE PRODUCERS

Hardware description.   This construct, shown in Figure 14.7(a), consists of a component B that merges two streams of data coming from components A_1 and A_2. These streams may have different rates. Component B reads values from its input channels according to its consumption pattern.

Model description.   The rate differences with multiple inputs causes some difficulty for a model in Petri nets. Because a Petri net transition can fire only if all of its input places contain at least one token, we cannot use a transition with multiple input places when we want to model a component that can read either one or two input values. A way to model this variable consumption behaviour is to create different transitions for each possible consumption pattern and a token that enables exactly one of these transitions. In our construct with two producers, we have three transitions as shown in Figure 14.7(b). Transition B models consumption of one token from component A_1, B' models consumption of two tokens from A_1 and A_2, and B'' models consump-

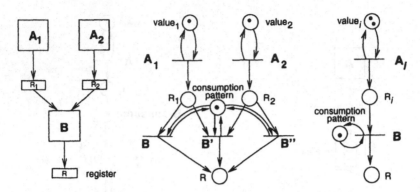

Figure 14.7 (a) Multiple producers, (b) Petri net model and (c) folded Petri net model.

tion of one token from A_2. The place *consumption-pattern* contains one token and is coloured to enable exactly one of the transitions B, B' or B'' that have preconditions to check the colour of this token. The consumption pattern can be represented as a list that is stored in the token in *consumption-pattern*. When the transitions B, B' and B'' consume a token, they place it back with the head of the list moved to the end, to enable the transition that is next in the consumption pattern.

Discussion.   The variable token consumption cannot be modelled in a straightforward manner, because we cannot define firing rules for the Petri net transitions like we can for hardware components. In Petri nets it is not possible to define a transition that consumes a variable number of tokens from its input places or that fires when not all input places contain tokens. A hardware component, on the other hand, can consume tokens even if not on all inputs data is available.

To model these kinds of variable rates, we had to split up the consumption and use consumption pattern tokens that indicate what token should be consumed next.

Folding

We can map the components A_1 and A_2 onto a single Petri net and apply token colouring to keep the tokens that are produced by components A_1 and A_2 separated. The resulting model is shown in Figure 14.7(c). We call this technique of modelling multiple hardware components of the same type by one single Petri net *folding*. This makes the model more compact and easily configurable for a variable number of hardware components, possibly with different parameters such as firing rates. Essential for this approach is the fact that transitions fire instantaneously. This makes it possible to model multiple instances of a component with one transition.

In our model of the hardware construct with multiple producers the folding technique reduces the number of transitions needed to model the possible consumption patterns. However, the delay of placing back the *consumption-pattern* token is different in the folded model. If component B consumes two input values simultaneously, transition B of the folded model consumes two tokens of different colour sequentially but instantaneously.

5.5   MULTIPLE CONSUMERS

Hardware description.   This construct, shown in Figure 14.8(a), consists of one component A that produces data for multiple consuming components B_i. Each clock cycle component A writes a value to register R. Components B_i read a value from register R at their own (possibly variable) rate.

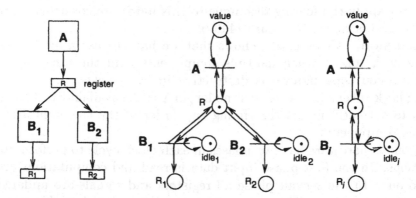

Figure 14.8   (a) Multiple consumers, (b) Petri net model and (c) folded Petri net model.

Model description.   The Petri net model of this hardware construct is shown in Figure 14.8(b). The value that component A writes to register R is inspected by more than one component. In Petri nets a value is read from a place by consuming the token and thereby removing the token from the place. Therefore the value can only be read once, unless it is placed back into the place from which it was consumed. An alternative approach is to let transition A produce a number of identical tokens such that there are enough tokens for all consumptions from the transitions B_i. This requires that transition A has information about the number of consuming transitions and their token consumption behaviour. Since we want to model the component A such that is independent of the number of consuming components and the rate at which they consume tokens, we consider this approach unacceptable. This means that we choose to let the each transition that models access to register R, place the token back into R to allow other transitions to read the value of this token as well.

A problem arises when transitions A and B_i are enabled simultaneously. When they fire, they will consume the token from place R and place it back. The transitions B_i will leave the value of the token unchanged, but transition A will change its value. Therefore the order in which the transitions fire is of importance. If we do not impose restrictions on the order in which transitions A and B_i can fire in our Petri net, our model will be non-deterministic.

This problem can be solved by extending the Petri net model with priorities associated with the transitions. By giving transition A a higher priority than transitions B_i, transition A will always fire first. As a result of this, all transitions B_i will read the new value of the token in place R.

We can apply the folding technique to this hardware construct. The folded model is shown in Figure 14.8(c).

Discussion.   This model shows that we have to use a method of placing back tokens when multiple components read the same input, because token consumption is destructive in Petri nets. A result of placing back tokens is that we can no longer use the availability of these tokens to correctly model the timing and order of the communication between components.

Synchronous hardware has two phases in each clock cycle to perform its operations. In the first phase input data is read and computations are carried out. In the second phase all registers and signals are updated with their new values. This ensures that the input data is stable when the output data is being computed.

In Petri nets computation of new output results and assignment of the results to places is not separated into phases. When a Petri net transition fires, it computes the output values and produces the results (possibly with a delay). When more than one transition is enabled at the same time, it is possible that each of the enabled transitions performs its operations on a different state of input values. Therefore we must impose explicit restrictions in our model such that transitions fire in the correct order. This can be done by using priorities for the transitions or using appropriate delays such that transitions will be enabled in the correct order.

6.   PETRI NET MODEL OF THE ARCHITECTURE

We have chosen to model the architecture at an abstract token level, that is, we define uninterpreted models for performance analysis. Tokens in our model represent samples of a video signal, but they do not contain the values of the samples. We do not model the functionality of the co-processors, but take only timing behaviour into consideration. This allows us to use our model for performance analysis. The accuracy of our performance analysis depends on the accuracy of the timing behaviour in our model. Because we do not model the functionality of the co-processors, we cannot cope with data dependent behaviour of co-processors. Data dependent behaviour must be approximated in our model with stochastic delays for the computation times and variable or stochastically determined consumption patterns.

We have used the structure of the Prophid architecture, shown in Figure 14.2, in our Petri net model of the architecture. This makes the model easy to understand because the flow of the tokens in the Petri net corresponds to a great extent with the flow of data in the architecture. We can model the components of the architecture using the models of the constructs presented in the previous section.

Because we want to be able to change the number of co-processors and the number of input and output buffers for each co-processor without having to make a lot changes in our model, we designed the model such that it provides this flexibility. We mapped all components of the same type onto one component that models them on separate levels. We call this the *folded* model of the architecture, shown in Figure 14.9(a).

The Petri net model of the architecture is shown in Figure 14.9(b). We use token colouring in our Petri net to separate the tokens of the different "levels" of our folded model. The label of a token indicates the location of the corresponding video sample in the architecture. We use timed Petri nets with delays associated with the production of tokens. The firing of a transition is instantaneous. This makes it possible to let one transition model different instances of a set of components, such as for example the input buffers.

This approach keeps the Petri net model compact and every possible instance of the class of Prophid architectures is captured with this Petri net. A configuration of the architecture with a certain number of co-processors and buffers is instantiated in the model with a set of token colours and an initial marking that represents the initial hardware state. For example, the set of buffers and their sizes in a configuration is deter-

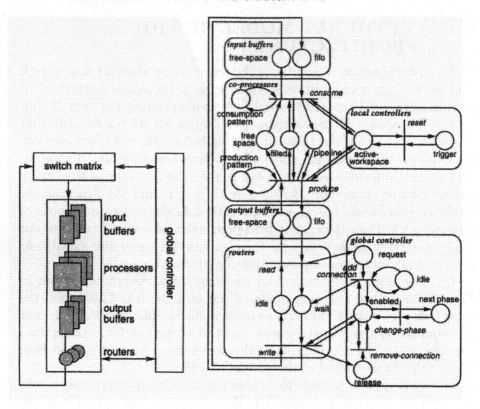

Figure 14.9 (a) Folded model of the architecture (b) Petri net model

mined by the number and colour of initial tokens in the *free-space* places of the buffers.

We applied the building block techniques that we presented in the previous section to construct the model of the architecture. The input and output buffers are modelled by two places representing buffer content and free space according to the FIFO buffer model. The co-processors consume video samples from a number of input buffers using some consumption pattern and have an internal pipeline for the computation. Therefore the *consume* transition that models consumption of video samples by the co-processors is a combination of the consuming transition of the FIFO buffer building block (Figure 14.5(b)), the consuming transition of the multiple producer building block with its *consumption-pattern* place (Figure 14.7(b)) and the consuming transitions of the pipeline building block (Figure 14.6(b)).

This also applies to the *produce* transition that models the production of video samples by the co-processors. Here the producing transitions from the same three building block models have been combined.

There are two places with multiple consumers in our Petri net model that require tokens to be placed back after they have been consumed. These involve the tokens residing in the place *current-workspace* (in *local-controller*) and the place *enabled* (in *global-controller*). Here we have to explicitly impose an order in the firing of transitions that consume a token from one of these places. This can be done by giving one transition priority over another.

The exact functionality of the *global controller* and *local controller* is beyond the scope of this paper. Therefore, we will not go into detail about the modelling of these components.

We measure the performance metrics of the architecture by extending the Petri net with transitions that collect tokens that mark an event in the architecture during execution of the application. We let the transitions that model the architecture, such as *consume*, produce a token each time a video sample token is consumed. These event tokens are translated to the appropriate value for the metric. For example, event tokens from the *consume* transition are translated into values for interarrival times and utilisation of the co-processor.

7.   SIMULATIONS

We used our model to simulate an instance of the Prophid architecture class with a benchmark application mapped onto it that imposes a realistic workload on the architecture. The simulations were carried out using ExSpect [1]. The architecture instance consists of 12 co-processors. The benchmark application is a video signal processing algorithm in the form of a dataflow graph consisting of 17 tasks, shown in a simplified form in Figure 14.10. The algorithm combines two video signals into a multi-window video signal. Two different input sources are used for the streams of video samples. A number of tasks are performed on the streams of video samples such as resizing and quality enhancement of the image before the two streams are combined into one output stream. The mapping that we used in the simulations is generated by a mapping tool. Different mappings are possible. The influence of different mappings on the performance can be evaluated by exercising the feedback loop to the mapping in the Y-chart (Figure 14.1).

The input and output signals consist of full size video frames. The video signals have a pattern of video data and blanking periods in between video frames. We ran a simulation for 4 video frames. The simulation time was approximately one day per video frame. This corresponds to nearly 1500 transition firings per second.

We measured the following performance metrics:

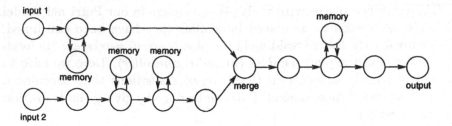

Figure 14.10   Benchmark application graph.

- utilisation of the co-processors and routers,

- buffer contents (average contents over time and relative frequency of different buffer fillings),

- throughput rates of the sample streams in the co-processors,

- response time for the global controller, and

- error rates for a number of real-time constraints.

Figure 14.11 shows some of the results of our simulations. In the measurements of the utilisation, shown in Figure 14.11(a), we can see the frame pattern of the sample stream. This graph shows the percentage of the total available processing time that one particular task of the dataflow graph requires from the co-processor onto which it has been mapped. In this case we see that it utilises 40% of the time during the active video frames and at the beginning of each frame it peaks at 50%. Figure 14.11(b) shows the average usage of an input buffer as function of the time. We see that during the active video frames the buffer of size 64 contains on average 48 samples. The plotted values are averages over intervals of 8192 clock cycles (2 video lines) in order to reduce the amount of measurement data. A disadvantage of this is that it evens out peaks in the measurement data. That is why we also measure the distribution of the number of times that each buffer usage occurs during the simulation. This is plotted in Figure 14.11(c). It shows that the same buffer after a run-in period always contains between 32 and 64 samples. These simulation results were used to verify that this co-processor never has underflow on its input buffer which was one of the real-time constraints.

8.   CONCLUSIONS

We have proposed a method to use timed coloured Petri nets to model dataflow architectures at a high level of abstraction. We identified five

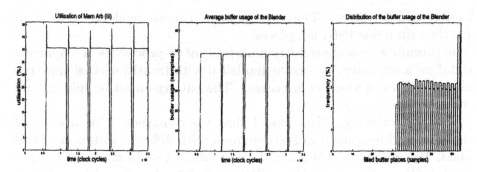

Figure 14.11 (a) Co-processor utilisation, (b) average buffer usage and (c) buffer usage distribution.

basic hardware building blocks that can be used to construct models of dataflow architectures. These five building blocks are sufficient to model the class of dataflow architectures of our interest. We used Petri net models of these building blocks to assess the value of Petri nets for modelling dataflow architectures.

We will draw our conclusions on the value of Petri nets as a modelling tool for performance analysis of dataflow architectures based the criterions that we presented in section 2..

Executability. Petri nets are executable, which enables us to simulate the modelled architecture. With simulations we can measure metrics that quantify the performance of the architecture under a workload that is imposed by a benchmark application and the streams of input data. We consider the availability of tools such as ExSpect and its ease of use to be important issues for this type of research.

Ease of modelling. Petri nets facilitate the modelling of parallelism in dataflow architectures. The timing behaviour of the hardware can be modelled with delays associated with tokens in the Petri net model. There is however a difference in the communication between parallel hardware components and Petri net transitions that can fire simultaneously, which deserves some attention when we model this. In hardware the computation of output data and updating of registers is separated into two phases. In Petri nets this is not separated. This can result in a possible state change of the modelled hardware after each firing. A sequence of transitions firing at the same moment in time to model parallel operations can have a different state for each firing. In hardware such intermediate state changes cannot occur. In some cases this requires additional Petri net constructs to explicitly impose a firing order of transitions to model the hardware behaviour correctly.

Petri nets offer expressive ways to model components such as FIFO buffers and pipelines. These components can be modelled very compactly with a few Petri net places.

The consumption and production patterns of co-processors in a dynamic dataflow architecture cannot be modelled with transitions that have the same firing rules as the co-processor. The patterns must be split up and modelled separately.

Configurability. The model that we constructed for a class of dataflow architectures can be configured with different instances of this class. We applied a *folding* technique to model a set of hardware components of the same type with one Petri net. This has the advantage that our model of the architecture can be configured without changing the Petri net. It is configured by defining the initial marking of the Petri net. By defining the initial marking we can also configure the model to execute a dataflow graph and mapping. This permits us to explore different points in the design space.

Efficiency. The main drawback of our approach is the efficiency of the simulations. Experiments showed that the efficiency was not very high for realistic applications involving large amounts of video samples. It takes a few days to simulate real life examples. The efficiency is determined by the way we modelled the architecture and by the speed of the simulation engine. The simulation engine that we used interprets an object code of the Petri net model. Higher simulation speed can be achieved with an executable program of the Petri net model which is specified in C++. This is currently under development.

Accuracy. We chose to model dataflow architectures at a high level of abstraction with only the timing and communication behaviour. The Petri net model does not contain the functionality of the co-processors or the exact hardware implementation of the components. Therefore our model is as accurate as the timing model that we use for the architecture. This involves assumptions about the computation time and stochastic approximations for possible data dependent behaviour. Given these abstractions in our model, the Petri net models produce accurate, cycle based performance numbers.

A specification of the Petri net model in C++ will facilitate to model the functionality of the co-processors. This will allow us to analyse data-dependent behaviour in the applications more accurately than by using stochastic approximations.

References

[1] ASPT (1994). *ExSpect 5.0 User Manual.* ASPT, Eindhoven University of Technology, Eindhoven, the Netherlands.

[2] Buck, J. (1993). *Scheduling Dynamic Dataflow Graphs with Bounded Memory Using the Token Flow Model.* PhD thesis, University of California, Berkeley.

[3] Govindarajan, R., Suciu, F., and Zuberek, W. (1997). Timed petri net model of multithreaded multiprocessor architectures. In *International Workshop on Petri Nets and Performance Models, IEEE*, pages 153–162.

[4] Jensen, K. (1992). *Coloured Petri Nets 1.* Springer Verlag, New York.

[5] Kienhuis, B., Deprettere, E., Vissers, K., and van der Wolf, P. (1997). An approach for quantitative analysis of application-specific dataflow architectures. In *Proceedings of 11th Int. Conference of Applications-specific Systems, Architectures and Processors*, pages 338–349.

[6] Leijten, J., van Meerbergen, J., Timmer, A., and Jess, J. (1997). Prophid, a data-driven multi-processor architecture for high-performance dsp. In *Proceedings of ED&TC,*.

[7] Lindemann, C. (1992). *Performance Analysis of Complex Systems by Deterministic and Stochastic Petri Net Models.* PhD thesis, Technische Universität Berlin.

[8] Madhukar, M., Leuze, M., and Dowdy, L. (1995). Petri net model of a dynamically partitioned multiprocessor system. In *International Workshop on Petri Nets and Performance Models, IEEE*.

[9] Marsan, M., Balbo, G., and Conte, G. (1986). *Performance Models of Multiprocessor Systems.* MIT Press, London.

[10] Molloy, M. (1982). Performance analysis using stochastic petri nets. In *IEEE Transactions on Computers, Vol. 31, No. 9*, pages 913–917.

[11] van Hee, K. (1994). *Information systems engineering: a formal approach.* Cambridge University Press, Cambridge.

[12] Watlington, J. and Bove, Jr., V. (1995). Stream-based computation and future television. In *Proceedings of the 137th SMPTE Technical Conference*, pages 69–79.

[13] Zuberek, W. (1991). Timed petri nets - definitions, properties and applications. In *Microelectronics and Reliability 31, No. 4*, pages 627–644.

Chapter 15

MODELING A MEMORY SUBSYSTEM WITH PETRI NETS: A CASE STUDY

Matthias Gries

Computer Engineering and Networks Laboratory (TIK),
Swiss Federal Institute of Technology (ETH) Zurich
CH-8092 Zürich, Switzerland
gries@tik.ee.ethz.ch

Abstract   Memory subsystems often turn out to be the main performance bottleneck in current computer systems. Nevertheless, the architectural features of common RAM chips are not utilized to their limits. Therefore, a complex Petri Net model of a memory subsystem was developed and investigated to explore possible improvements. This chapter reports the results of a case study in which widely used synchronous RAMs were examined. It demonstrates how impressive throughput increases can be obtained by an enhanced memory controller scheme that could even make second level caches redundant in cost or power dissipation critical systems.

Furthermore, using colored time Petri nets leads to a descriptive view of the memory subsystem because a Petri Net model combines data and control flow as well as structural information in a natural way. The case study finally underpins the advantages of this approach.

Keywords: Colored Petri nets, instruction scheduling, memory systems, modeling

1.   INTRODUCTION

The memory subsystem (see Fig. 15.1) in a computer usually consists of a controller and several memory chips. The controller receives read and write requests from the central processing unit, modifies them, and generates additional instructions for the memory chips if necessary. Common asynchronous memory chips like Extended Data Output (EDO) [15] chips are not able to process several instructions concurrently. Thus, common controllers schedule requests from the CPU

A. Yakovlev et al.(eds.), Hardware Design and Petri Nets, 291-310.
© 2000 *Kluwer Academic Publishers.*

sequentially for the memory chips. However, recent trends in memory architectures show an evolution towards concurrent structures, e.g. synchronous DRAM chips (SDRAMs [15]). Ordinary controllers supporting EDO and SDRAM chips do not distinguish between these two types of memory. Therefore, SDRAMs are rarely utilized to their full potential, that is, they are not capable of obtaining higher throughput rates than asynchronous chips. At the same time, clock rates and throughput demands of modern CPUs grow constantly [2].

The computer manufacturing industry reacts by increasing the clock frequency of memory interfaces (not the memory array itself) [13]. Moreover, memory array sizes can be reduced in order to decrease RAM access delays. Finally, more cache levels can be introduced in the memory hierarchy but the costs for cache memories are not negligible.

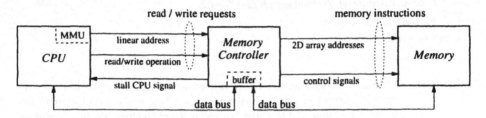

Figure 15.1   Generic view of a computer system.

Nevertheless, a major part of wasted memory throughput is explained by the fact that memory controllers are still not able to take advantage of concurrently operating structures within memory chips. The case study therefore indicates the potential in terms of throughput increases by modifying the memory controller instruction schedules.

Requirements for a modeling tool

The design of complex computer systems usually involves people from different fields of expertise, e.g. control and data flow dominant parts of the system could be developed by separate teams in parallel. Accordingly, the following requirements for a suitable modeling tool can be deduced:

- Support for structuring the model into modules and components and for reusability has to be offered.

- The process of modeling must be supported by a natural representation of control and data flow as well as structural properties.

- Simulation and documentation (e.g. getting statistics data out of a simulation run) of the system must be provided.

- Configurability and parameterization of modules and components should be possible.

- Modeling of concurrency and a notion of time are mandatory for the detection of timing violations, for performance evaluation, and for verification of run-time behavior.

Modeling alternatives

State machine based approaches, such as StateCharts [9] or ROOM-chart [17], are best suited for control dominated systems and suffer from their inability to express data flow. Another discrepancy between a system and its model representation can be found looking at all the tools that do not allow to express structural similarity between a system and its model, e.g. spreadsheet based models and models written exclusively in the form of programming language code. Recently, the use of object-oriented modeling [3, 16] becomes more and more common. Although object-oriented formalisms contain several features to produce detailed models, they are not intended to be executable as such. Place/Transition Petri nets [14] have several desirable properties, such as being intuitive, graphical, able to express concurrency and data flow. However, they are confined to the use in small scale models because a concept of hierarchy is missing. High-level Petri nets (such as Coloured Petri Nets [12]) are better suited, since they have an expressive inscription language and also some structuring features.

In the next section, a short overview of present memory systems is given including currently available cache and main memory technologies. In section 3., the main features of our Petri net modeling and simulation tool are outlined and checked against the mentioned requirements. In the remaining part of the chapter, section 4., the modeled memory system is presented in detail.

2.   MEMORY CHARACTERISTICS
2.1   COMMONLY USED MEMORY CHIPS

Presently, asynchronous memory chips like EDO RAMs are found in a variety of personal computers, printer devices, workstations, digital assistants, etc. and hence dominate the memory market. The term *asynchronous* describes the fact that a memory controller has to hold control signals for the memory array on the same voltage during the whole memory transfer. In consequence, the controller cannot issue another memory instruction unless the previous one is finished. That way, even though several memory banks are available on a single chip, the controller is forced to schedule memory instructions strictly sequentially since only one set of control signals can be accessed by the controller at a time.

2.2   SYNCHRONOUS MEMORIES

Synchronous DRAMs Recent RAMs like SDRAM or RDRAM [7] are enhanced versions of asynchronous RAMs. They additionally have a synchronous interface that isolates the main memory cell array from the signals of the memory controller. As the interface has a separate command pipeline for every memory bank, the controller now has the option to issue memory instructions on each clock cycle. Undoubtedly, it is normally not advisable to transfer a memory instruction on each clock cycle since parallel memory banks have to share a single data bus and an input/output buffer pair. But a memory bank can be prepared for a data transfer while another bank is actually transferring data concurrently. This preparation is required due to address decoding and memory access times as well as the fact that dynamic memory cells must be precharged after they were accessed. However, not all delays can be hidden completely by using parallel memory banks concurrently, for instance, when the direction of data flow through the input/output buffers changes.

Moreover, if timing constraints of the memory chips (e.g. pipeline and memory array timings) are not met when transferring instructions, there is a danger of data contention inside the memory arrays because memory chips do not check against forbidden control sequences.

Finally, current controllers schedule memory instructions in almost the same sequential manner for synchronous RAMs as for asynchronous ones.

Second level caches Typical second level caches and synchronous DRAMs exhibit almost identical behavior. They are both synchronous and pipelined and can be programmed to burst transfers, that is, data on successive addresses are transferred without the help of the memory controller. In systems where external caches run at the same cycle frequency as the main memory (as in most currently used personal computers), the speed difference is less than a factor of two (see section 4.4 for details). Even though caches are still a little bit faster because their charge does not need to be refreshed using static RAM technology and their memory bank size is smaller, an optimized memory controller could compensate the need for second level caches in power dissipation or cost critical systems.

3.   THE MODELING ENVIRONMENT

In this section, the CodeSign tool is presented and it is pointed out how the requirements for a modeling tool are met. A more detailed description can be found in [8]. CodeSign is freely available [6]. CodeSign is based on a kind of colored Petri nets that allow efficient modeling of control and data flow.

Components, composition, and hierarchy Components are subnets of Petri nets with input and output interfaces that are applied to interconnect components. Inside components, input interfaces are connected to places and transitions are connected to output interfaces. Linking output with input interfaces, components are directly interconnected maintaining Petri Net semantics. In Fig. 15.2 the model of a RAM basic cell with its interfaces is shown as an example. The model is explained in detail in section 4.1. If input interfaces of components

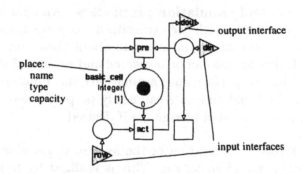

Figure 15.2   Petri Net model of a RAM cell component.

are connected together, a single token will produce several tokens with the same data value for each component. Connecting output interfaces together is the same as connecting several transitions to a single place. Therefore, the examples in Fig. 15.3 are equivalent.

Thus, models can be hierarchically structured and verified components can be reused. Moreover, as interfaces do not disturb Petri net semantics, a flat Petri Net with the same functionality can always be generated from the hierarchically structured net.

Object oriented concepts Components are instances of classes that are arranged within an inheritance tree. That is, classes inherit features and (token) data types from their parents. They may contain functions which can be used in transitions. Functions are written in an appropriate imperative language. With these facilities, incremental updates and evolution of models and components are supported as well as configurability and parameterization.

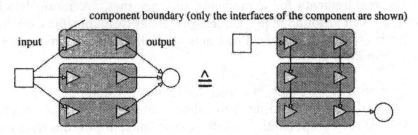

Assumption: the transition on the left produces tokens of the same type for each component.

Figure 15.3 Using interfaces in CodeSign.

Notion of time and simulation properties   An enabling delay interval can be associated with each transition. As a consequence, tokens carry time stamps containing information about their creation date.

Models and components can be inspected and simulated at all levels of abstraction. That is, performance evaluations and functionality checks like race conditions and timeouts can easily be performed. Finally, the simulation can be animated at runtime if desired.

Instrumentation   Observation of the system is possible without affecting the main model structure. This is realized by introducing so-called *probe* elements which collect statistical data like firing times of transitions or generation dates of certain tokens. This information can be stored or further processed by Petri Net components. Probes are invisible in the main model because they are defined in a special editor. Bindings associate probes with transition or place elements.

4.　PETRI NET MODEL OF A MEMORY SUBSYSTEM

In this section, the modeled memory subsystem is described which includes a memory controller, a synchronous memory chip as well as the relevant parts of the CPU and the data bus. In Fig. 15.4 an overview of the main system structure is given.

The CPU issues *data read* and *data write* requests. They are modeled as tokens which contain a list of values. Each request token carries information about the type of the request (read or write) and the address (an integer value) where the data can be found in memory. The sequence

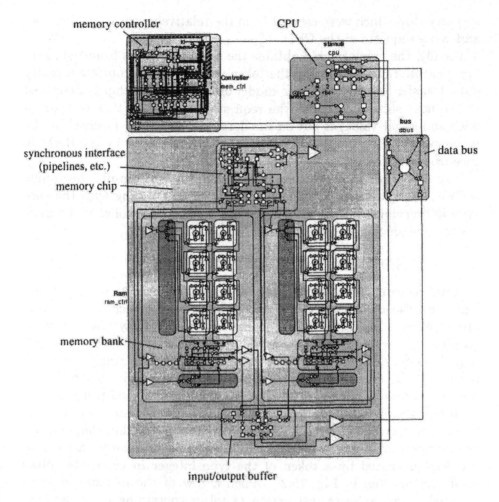

memory controller

CPU

synchronous interface
(pipelines, etc.)

memory chip

data bus

memory bank

input/output buffer

Figure 15.4 CodeSign screen shot showing the model of a memory subsystem.

of requests can be arbitrarily chosen and may be extracted from address traces generated by a tool like [5].

After having received a request token from the CPU, the memory controller preprocesses the token for the SDRAM chip. The address value of the token has to be split into memory bank, row, and column addresses as these values must be transferred at different times to the RAM chip which is organized as several two-dimensional memory arrays. Besides, additional instruction tokens are generated which are necessary in order to prepare the memory, in particular tokens that initiate activate or precharge behavior of the memory cells. The memory controller is explained in more detail in section 4.2. At this stage, the instruction tokens may be seen as micro instructions for the control logic of the memory chip which were created from the relatively coarse grained read and write requests of the CPU.

Finally, the memory chip obtains the modified tokens from the memory controller. Depending on the token values, the appropriate memory data transfer is initiated. The model of the memory chip is described in the next section. At last, the requested data item (that is, a token with an integer value) is put on the data bus and either received by the memory controller in case of a read request or received by the RAM in case of a write request.

The explanation of the model is organized "bottom-up" beginning with a description of the memory chip itself and ending with the subsystem overview. The description of the memory controller is the most precise one since it forms the core of the whole model.

4.1   SDRAM ARCHITECTURE

Asynchronous RAM array   As already stated in section 2.2, the core logic and the memory array of synchronous and asynchronous RAMs are identical. In Fig. 15.5, a conventional RAM array with its sense amplifiers, row, and column decoders can be seen (a *memory bank*). The decoders are necessary to access the contents of a single RAM cell in the two-dimensional array of cells. A single cell in general holds four to 16 bits of information. The decoders are realized using mutual exclusive guard functions within the transitions because only a single transition may be enabled per token. Token values within the decoders are interpreted as row or column addresses. The data within a memory cell is represented by a token of the type integer in the center place as it can be seen in Fig. 15.2. When a row of the memory array is selected by an *activate* instruction (a token containing a row address value, a memory bank address, and an operation identifier) issued by the memory controller, the information of that array row, that is, all data tokens in a row of memory cells, is transferred into the so-called *sense amplifiers*. Thus, the corresponding transition in the row decoder fires and all memory cells of that row are activated through their *row* input interfaces. As it can be seen in Fig. 15.2, the transition *act* is enabled and finally the information of the memory cell (the integer token in the center place) is transferred via the output interface *dout* to the sense amplifiers.

Sense amplifiers   The sense amplifiers are necessary since the electrical charge of dynamic memory cells is too weak to be used directly. After activating a memory row, the data tokens in the sense amplifiers can be read or updated column by column depending on the current

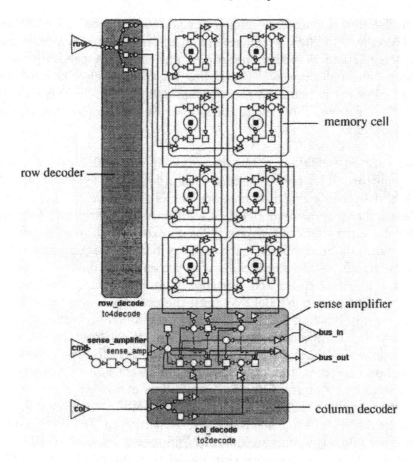

Figure 15.5 A conventional asynchronous RAM array with $4 \times 2 = 8$ memory cells.

mode of operation (read or write). The read or write state is reached when the memory controller issues *read* or *write* instruction tokens each containing an operation identifier and a column address.

Read and write operations are destructive, that is, once the information is transferred into the sense amplifiers, the charge in the memory cells is lost. Therefore, the information in the sense amplifiers must be written back into the memory cells when the data of that special row is no longer needed. This type of operation is forced by a *precharge* command of the memory controller. In this case, the integer tokens carrying the memory information are written back to the memory cells through the *din* input interface of the memory cell. The transition *pre* is enabled in case the corresponding row is decoded and the possibly modified data token is returned to the center place of the memory cell.

All phases of a complete memory cycle may be seen in an animated simulation in CodeSign. It consists of an activation of a specific memory row, several read or write operations, and finally a precharge of the memory cells in that row. Depending on the modeled memory chip, the different operations require variable delays, e.g. in this case [10], an activation takes three clock cycles distributed through decoders and sense amplifiers.

Synchronous interface   The memory chip is completed by adding an input/ output buffer pair, registers, control logic (the synchronous interface), and by using multiple memory banks that have to share buffers. Buffers and registers are modeled as additional places with limited capacity for data or instruction tokens. The control logic implements several instruction pipelines, one for each memory bank. The pipelines are represented by successive transitions with an enabling delay of one clock cycle and places with limited capacity. That way, tokens carrying operations and addresses are delayed until the corresponding asynchronous memory array has finished the previous operation. Furthermore, the address may be incremented for burst transfers of consecutive addresses.

Thus, memory banks are able to work concurrently due to separated instruction pipelines, but only a single one is actually able to transfer data through the shared input/output buffers at a time.

In the CodeSign model, four timing parameters of the modeled synchronous RAM [10] are considered, e.g. activation and precharge delays. All modeled delays within an SDRAM component are derived from these constants. Thus, the memory chip model (a component class in Code-Sign) can quickly be adapted to SDRAMs from other manufacturers or even to other memory technologies.

4.2   MEMORY CONTROLLER

Functionality   A memory controller receives requests from the CPU at arbitrary times. They consist of the type of operation (read or write) and address information (typically 32 bit addresses for a bytewise linearly addressable memory space). Sometimes, this address information has been preprocessed by a memory management unit (MMU, see Fig. 15.1) that usually maps addresses to other memory regions due to memory page limits or protected memory regions (e.g. reserved for memory mapped IO). The memory controller now has to take care of delays of the memory array according to the corresponding data sheet. Since there are typically several memory banks within a memory chip, the controller must prevent data tokens from collision with other tokens e.g. on the data bus. That is, the controller has to transfer instruction tokens to the memory chip at points of time which are sufficiently apart from each other because the memory chip will translate the instructions into actions on the memory cells without any timing checks.

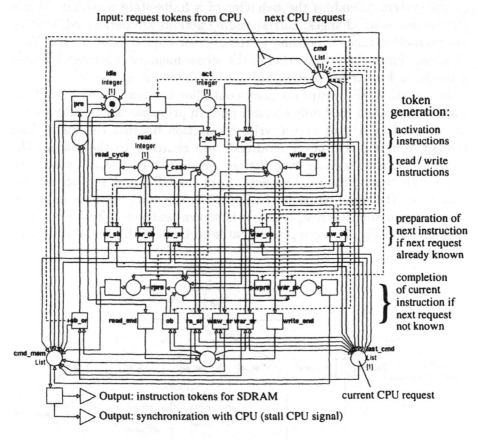

Figure 15.6   Model of the memory controller.

Petri Net model   At first, the controller has to split the address information of the request tokens from the CPU into several tokens since memory chips are organized as two-dimensional arrays. The split address information is now transferred to the selected memory chip at different times within the memory access cycle. That is, the respective memory row must be opened with an *activate* command (token) which consists of the corresponding instruction identifier and the bank and row addresses. Then, the request token itself is transmitted to the memory chip which consists of an instruction identifier of a read or a write operation and the bank and column addresses. Read and write operations on the same memory row can be repeated several times until finally the memory row must be closed. For this, the controller transfers a *precharge* instruction token holding an identifier for this type of operation as well as bank and row addresses. This control flow dominant part of the system resembles the behavior of a finite-state machine. When the current state changes, corresponding actions are performed, that is, an instruction token according to the current request token of the CPU is issued. Since the explanation of the whole memory controller requires a profound knowledge of SDRAMs, it is left out here.

In Fig. 15.6, the entire memory controller is shown. The dotted arrows are so-called *read only* connections. In principle, they could always be replaced by a read and a write connection because the token data is read from a place, never modified, and returned to the place. The controller can be seen as two distinct Mealy-like finite-state machines, one for processing the read requests from the CPU and another one for processing the write requests. In Fig. 15.7, the state machine responsible for processing the read requests is shown (extracted from Fig 15.6). Not shown in Fig. 15.7 are places that keep track of the global state (the

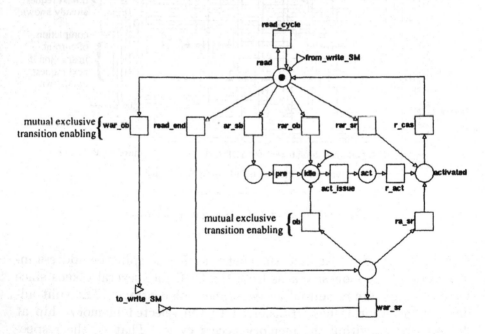

Figure 15.7   State machine handling the read requests.

request token currently in process) or that buffer request and instruction tokens at the interfaces of the controller. In other words, Fig. 15.7 shows only the state transitions and not the input and output places which are responsible for the mutual exclusive transition enabling.

Let us assume that a read request is currently in process, as shown in Fig. 15.7, and another read request which is already known in advance should be processed next. This next read should access another memory row of the same memory bank. Thus, the transition ar_sb is enabled (ar_sb stands for "read after read, same bank"). As an example of a guard function, the condition for enabling the transition ar_sb is shown in Fig. 15.8. The function checks the different value fields of the current tokens in the places *cmd*, *last_cmd*, and *read*. The places *cmd* and *last_cmd* are displayed in the top-right and the bottom-right corners of Fig. 15.6. The transition ar_sb fires after a certain amount of

(cmd[1] = 'read') & (cmd[2] = last_cmd[2]) & (cmd[3] <> last_cmd[3]) & (read >= (burstl() - 1))

check if the next request is another read request	check if the current and the next request want to access the same memory bank	check if the current and the next request want to access different memory rows	check if the current read request has been processed completely (the function burstl() returns the burst length)

Figure 15.8 Guard function of the transition ar_sb.

time which depends on the used SDRAM and the chosen burst transfer mode. This firing generates an instruction token for the SDRAM that starts the precharge of the corresponding memory row. Then, the transition *pre* is enabled and fires after the SDRAM precharge time constant. Therefore, when reaching the place *idle* the corresponding memory bank will be in the idle state (the memory bank is precharged) and the controller can issue another instruction token for the SDRAM by firing the transition *act_issue*. This instruction token activates the corresponding memory row for the next read request. The transition r_act fires after the SDRAM activation delay and generates a read instruction token for the SDRAM. Finally, the transition r_cas models the constant delay between the issue of the read instruction and the appearance of the first data item on the data bus.

In Fig 15.9, a black box view of the controller is given in which three arbitrary read request tokens from the CPU produce six instruction tokens on the output interface of the memory controller. All tokens for a whole memory cycle containing an activation, two reads, and a precharge can be seen followed by another activation and a read instruction. The time stamps in this example are integer values as requests and instructions are transferred at multiples of the clock cycle time.

The read-after-read access explained above can be observed between the second and third request token in Fig 15.9. The read-after-read request for an entry in the same memory bank but different memory row

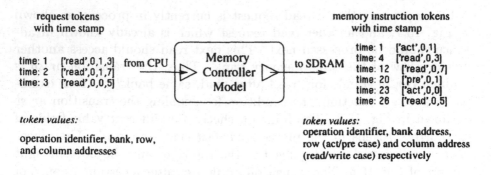

Figure 15.9 Instruction token generation by the memory controller.

causes the generation of two additional instruction tokens for precharge and activation. Moreover, the issue of some tokens must be delayed in order to cope with the timing restrictions of the SDRAM.

Model characteristics Basically, the controller behaves like commonly used controllers in PCs, i.e. it schedules read and write requests of the CPU sequentially without affecting their order. However, it is capable of abbreviating the latency penalty which is usually introduced by a change of the memory row or bank in case the next request by the CPU is already known. Thus, additional distinctions of cases which consider the previous instruction more precisely and the fine adjustment of enabling delays of transitions are the core enhancements in comparison with common memory controllers.

In the CodeSign model, six timing parameters of the modeled controller are considered that depend on the memory used, that is, they are a property of the according class. Thus, the memory controller model can be quickly adapted to other memory standards or memories from other manufacturers.

The flow of instructions from the controller to the RAM is modeled using tokens of type *list* which consist of a string value and two integer values (see Fig. 15.9 on the right). This way, the tokens emulate commands with an operation field (read, write, precharge, activate) and two address fields (bank address, row and column address respectively). Besides, a token with an integer type is used to model the current state of the controller. The value is used as relative time reference in clock cycles during the whole memory cycle of the corresponding read or write request.

4.3 MEMORY SUBSYSTEM

The memory subsystem is completely modeled by adding a data bus and a CPU. The data bus is shared for read and write data and therefore realized by a mutual exclusive access scheme. Using a random number generator in CodeSign, the CPU transfers read and write request tokens at arbitrary times if desired. These tokens may be extracted from previously generated address traces. The flow of read and write requests from the CPU to the memory controller is modeled using tokens of type *list* which consist of a string value and three integer values (see Fig. 15.9 on the left). Thus, the tokens may be seen as instructions with an operation identifier (read or write request) and one big address field which is already grouped into three main areas (bank, row, and column address) e.g. by a memory management unit. In addition, the controller may stall the CPU in order to synchronize data transfers between CPU and RAM.

4.4 RESULTS

The complete model consists of 114 places, 140 transitions, 535 connections, and 12 probes and was created by the author with only little prior experience of Petri Nets in a period of approximately three weeks. A kind of high level Petri Nets was chosen as modeling formalism instead of Place/Transition nets. Since the main focus of the investigations were performance issues as the determination of the data bus usage and throughput as well as the duration of schedules, the possibility to associate periods with transitions in timed Petri Nets was preferred to the option to formally check properties like liveness or reachability on a Place/Transition net in all details. Furthermore, it was a great advantage to use colored tokens and guard functions because the size of the whole net became sufficiently small by shifting some complexity into transitions and tokens to facilitate a quick overview of the entire system. We thus employed a similar methodology as the one presented in [4] in which by modeling the instruction execution of a microprocessor the same problems were dealt with as for instance data-dependent conditions of action execution and accurate instruction flow representation.

Finally, the similarity between the hardware system and its model was additionally supported by (hierarchically) using subnets as components.

Model analysis with CodeSign In particular, the development and improvement of the model has been accelerated by the instrumentation and graphical simulation features of CodeSign. Several race conditions have been found quickly as well as architectural faults of the model. See

Fig. 15.10 for an example. Frequently, wrong schedules of the memory controller such as missed out instructions or too dense instruction issues are the main cause for erroneous conditions.

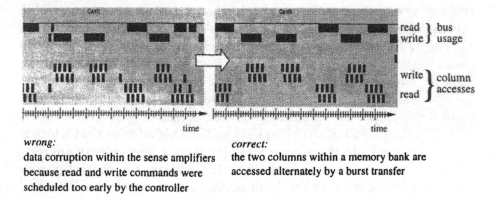

wrong:
data corruption within the sense amplifiers because read and write commands were scheduled too early by the controller

correct:
the two columns within a memory bank are accessed alternately by a burst transfer

Figure 15.10   Instrumentation: The firing times of transitions were collected.

Model development and verification   The model of the memory chip was developed first. It was checked with the help of fixed, previously composed memory instruction sequences for which the default behavior of the memory chip was already known. After having adjusted all transition enabling delays according to the corresponding data sheet [10], the memory controller and its surrounding components (bus, CPU) were added. Again, previously composed sequences of instructions were used, in this case sequences of read and write request tokens, in order to check the model. Enabling delays were minimized in order to be able to schedule the next memory instruction as early as possible but without causing any deadlocks within the memory chip.

Performance comparison   The shortened schedules of memory instructions issued by the memory controller in comparison with conventional controllers have been realized by about ten additional transitions and several places as well as the fine tuning of enabling delays of transitions within the controller.

In Tab. 15.1 the memory access delays of some popular memory controllers for a four times burst transfer are shown. That is, four column entries in a memory array row are transferred consecutively. This is the typical mode of operation in common PCs. The access types shown are the most frequently found access schemes in address traces. Altogether, 14 different access types have been considered in the study. For instance, a read operation after a read operation on another memory row of the

Table 15.1 Speed comparison between popular controllers. The accesses are performed in four times burst mode. The delays are given in *clock cycles*[a].

SDRAM access type		memory controller access delay		
		Intel [11]	*SiS [18]*	*AMD [1]*
read	row hit	7-1-1-1	7-1-1-1	7-1-1-1
	row start	9-1-1-1	10-1-1-1	9-1-1-1
	row miss	12-1-1-1	13-1-1-1	11-1-1-1
	b-b[b] burst	2-1-1-1	3-1-1-1	3-1-1-1
write	row hit	3-1-1-1	3-1-1-1	not
	row start	6-1-1-1	6-1-1-1	specified
	row miss	9-1-1-1	9-1-1-1	

[a] 66 MHz clock, $t_{CAS} = 3$ cycles, see [10] for details.
[b] back to back burst read, row hit.

same memory bank is called a *read row miss* because the same memory row cannot be used for both operations. Otherwise, the second operation would be called a *read row hit*. The delays in clock cycles are given in the table for each memory column entry. An expression like *7-1-1-1* means that the first entry needs seven clock cycles until it is delivered to the memory controller in case of a read request or saved into the sense amplifiers in case of a write request. The next three entries only need one cycle each due to pipelined transfers. Therefore, the whole burst transfer needs ten clock cycles in this example.

In Tab. 15.2 the memory access delays of the CodeSign memory controller model are given. The potential performance improvements by using this controller in comparison to the ones given in Tab. 15.1 are shown for the corresponding access type in the last two columns of Tab. 15.2. It can be seen that the enhanced controller modeled in Code-Sign is almost always faster than the quoted common controllers. The performance gain has been mainly achieved by optimized, compressed command schedules of the controller in case the next read or write requirement of the CPU is already known. In particular, the property of preparing a data transfer of one memory bank and transferring data with another bank concurrently has been exploited.

All quoted popular controllers also control the external second level cache which runs at the same clock speed as the main memory in the computer system. In Tab. 15.3 the delays for cache accesses are compiled. Looking at Tab. 15.2 the delays of the CodeSign model are close to the access delays for external caches. Therefore, cost or power dis-

Table 15.2   Speed comparison between the popular controllers given in Tab. 15.1 and the CodeSign controller model. The accesses are performed in four times burst mode. The delays are given in *clock cycles*[a].

SDRAM access type		CodeSign model max.[c]	min.[d]	speed increase min.	max.
read	row hit	7-1-1-1	1-1-1-1	0%	60%
	row start	7-1-1-1	1-1-1-1	17%	69%
	row miss	10-1-1-1	7-1-1-1	7%	33%
	b-b[b]) burst	3-1-1-1	1-1-1-1	0%	33%
write	row hit	1-1-1-1	1-1-1-1	33%	33%
	row start	4-1-1-1	1-1-1-1	22%	56%
	row miss	7-1-1-1	4-1-1-1	17%	42%

[a] 66 MHz clock, $t_{CAS} = 3$ cycles, see [10] for details.
[b] back to back burst read, row hit.
[c] Maximum delay: The next request by the CPU is not known in advance.
[d] Minimum delay: The next request is known while processing of the current one.

sipation critical embedded systems may drop the support for external caches in favor of an optimized memory controller design.

That is, with the help of the quickly developed model, promising performance increases were deduced. Further investigations may focus on examinations of address traces of real applications in order to quantify how often the different address types are actually used. Accordingly, virtual performance improvements can be determined which depend on the application analyzed and the properties of the memory used in the computer system.

Table 15.3   Speed comparison of second level cache accesses in four times burst mode. The delays are given in *clock cycles* of a 66 MHz clock.

2. level cache access type	memory controller access delay Intel [11]	SiS [18]	AMD [1]
arbitrary read / write	3-1-1-1	3-1-1-1	3-1-1-1
b-b[a] burst	1-1-1-1	1-1-1-1	1-1-1-1

[a] back to back burst read or write.

5. CONCLUSION

In this chapter, a complex model of a memory subsystem of a computer system was shown. With its help remarkable speed increases with small architecture enhancements were realized in comparison with common solutions. The improvement was derived by using parallel memory banks in memory chips concurrently. The complete model with optimizations was created within a short period with the help of the CodeSign Petri Net modeling environment which is based on colored time Petri nets and which fulfills all necessary requirements in order to represent control and data flow as well as structure properties of real systems in an intuitive way.

The case study showed that the functionality of memory controllers should be improved primarily as RAM array delays remain the same and only the memory interface speed increases. Furthermore, an optimized controller may be considered as an alternative to second level caches in power dissipation or cost critical systems.

The model is easily adaptable to new standards and technologies using its configurability and the inheritance feature of the CodeSign environment.

Acknowledgments

The author would like to thank R. Esser, J. Janneck, M. Naedele, and L. Thiele for valuable comments and discussions.

References

[1] Advanced Micro Devices Inc. *AMD-640 System Controller*, June 1997.

[2] Keith Boland and Apostolos Dollas. Predicting and precluding problems with memory latency. *IEEE Micro*, 14(4):59–67, August 1994.

[3] Grady Booch. *Object-oriented analysis and design, with applications*. Benjamin/Cummings, 2nd edition, 1994.

[4] Frank P. Burns, Albert M. Koelmans, and Alex V. Yakovlev. Analysing superscalar processor architectures with coloured Petri nets. *International Journal on Software Tools for Technology Transfer, Springer-Verlag*, 2(2):182–191, 1998.

[5] Bob Cmelik and David Keppel. Shade: A fast instruction-set simulator for execution profiling. *Proceedings of the 1994 ACM SIGMETRICS Conference on the Measurement and Modeling of Computer Systems*, pages 128–137, May 1994.

[6] Computer Engineering and Networks Laboratory (TIK), Swiss Federal Institute of Technology (ETH) Zurich, Switzerland, *http://www.tik.ee.ethz.ch/~codesign. CodeSign project.*

[7] Richard Crisp. Direct Rambus technology: The new main memory standard. *IEEE Micro*, 17(6):18–28, November 1997.

[8] Robert Esser. *An Object Oriented Petri Net Approach to Embedded System Design.* PhD thesis, Computer Engineering and Networks Laboratory (TIK), ETH Zurich, Switzerland, 1996.

[9] David Harel. Statecharts: A visual formalism for complex systems. *Science of Computer Programming*, 8(3):231–274, 1987.

[10] IBM Corp. *64Mb Synchronous DRAM - Die Revision A, IBM0364164C*, March 1998.

[11] Intel Corp. *430TX PCISET: 82439TX System Controller (MTXC) Specification & Specification Update*, March 1998.

[12] Kurt Jensen. *Coloured Petri Nets: Basic Concepts, Analysis Methods and Practical Use*, volume 1: Basic Concepts of *EATCS Monographs in Computer Science*. Springer-Verlag, 1992.

[13] Yasunao Katayama. Trends in semiconductor memories. *IEEE Micro*, 17(6):10–17, November 1997.

[14] Tadao Murata. Petri nets: Properties, analysis, and applications. *Proceedings of the IEEE*, 77(4):541–580, April 1989.

[15] Betty Prince. *High Performance Memories.* John Wiley & Sons Ltd., 1996.

[16] James Rumbaugh, Michael Blaha, William Premerlani, and Frederick Eddy. *Object-oriented modeling and design.* Prentice Hall, Englewood Cliffs, New Jersey, USA, 1991.

[17] Bran Selic, Garth Gullekson, and Paul Ward. *Real-time object-oriented modeling.* John Wiley & Sons Ltd., 1994.

[18] Silicon Integrated Systems Corp. *SiS5591/5592 Pentium PCI A.G.P Chipset*, January 1998.

Chapter 16

PERFORMANCE MODELING OF MULTITHREADED DISTRIBUTED MEMORY ARCHITECTURES

Wlodek M. Zuberek
Department of Computer Science
Memorial University of Newfoundland
St.John's, NF, Canada A1B 3X5
wlodek@cs.mun.ca

Abstract   In multithreaded distributed memory architectures, long–latency memory operations and synchronization delays are tolerated by suspending the execution of the current thread and switching to another thread, which is executed concurrently with the long–latency operation of the suspended thread. Timed Petri nets are used to model several multithreaded architectures at the instruction and thread levels. Model evaluation results are presented to illustrate the influence of different model parameters on the performance of the system.

Keywords: distributed memory architectures, multithreaded architectures, performance modeling, timed Petri nets

1.   INTRODUCTION

The performance of microprocessors has been steadily improving over the last two decades, doubling every 18 months. On the other hand, the memory performance and the performance of the processor interconnecting networks have not improved nearly as fast. The mismatch of performance between the processor and the memory subsystem significantly impacts the overall performance of both uniprocessor and multiprocessor systems. Recent studies have shown that the number of processor cycles required to access main memory doubles approximately every six years [4]. This growing gap between the processing power of modern microprocessors and the memory latency significantly limits the performance

A. Yakovlev et al.(eds.), Hardware Design and Petri Nets, 311-331.
© 2000 Kluwer Academic Publishers.

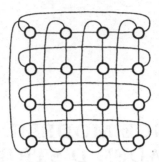

Figure 16.1 Outline of a 16–processor system.

of each node in multicomputer systems. Trends in processor technology and memory technology indicate that this gap will continue to widen at least for the next several years. It is not unusual to find that the processor is stalled 60% of time waiting for the completion of memory operations [13].

In distributed memory systems, the latency of memory accesses is much more pronounced as memory access requests may need to be forwarded through several intermediate nodes before they reach their destination, and then the results need to be sent back to the original nodes. Each of the 'hops' introduces some delay, typically assigned to the switches that control the traffic between the nodes.

Instruction reordering is one of the approaches used to alleviate the problem of divergent processor and memory performances. Multithreading is another approach which combines software (compilers) and hardware (multiple thread contexts) means [1, 5, 15].

Multithreading is an architectural approach to tolerating long–latency memory accesses and synchronization delays in distributed memory systems. The general idea is quite straightforward. When a long–latency memory operation occurs, the processor instead of waiting for its completion (which in distributed memory systems can easily require a hundred or more processor cycles), switches to another thread if such a thread is ready for execution. Switching to another thread can be performed very efficiently because the threads are executing in the same address space. If different threads are associated with different sets of processor registers, switching from one thread to another can be done in one or just a few processor cycles [6, 7].

A distributed memory system with 16 processors connected by a 2–dimensional torus–like network is used as a running example in this paper; an outline of such a system is shown in Fig.16.1.

It is usually assumed that the messages sent from one node to another are routed along the shortest paths. It is also assumed that this routing is done in a nondeterministic way, i.e., if there are several shortests paths

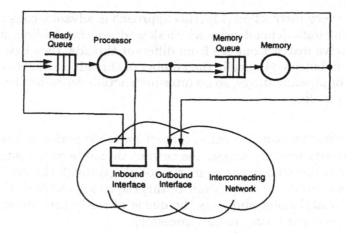

Figure 16.2 Outline of a single multithreaded processor.

between two nodes, each of them is equally likely to be used. The average length of the shortest path between two nodes, or the average number of hops (from one node to another) that a message must perform to reach its destination, is usually determined assuming that the memory accesses are uniformly distributed over the nodes of the system.

Although many specific details refer to this 16–processor system, most of them can easily be adjusted to other systems by changing the values of a few parameters. Some of these adjustments are discussed in the concluding remarks.

Each node in the network shown in Fig. 16.1 is a multithreaded processor which contains a processor, local memory, and two network interfaces, as shown in Fig. 16.2. The outbound switch handles outgoing traffic, i.e., requests to remote memories originating at this node as well as results of remote accesses to the memory at this node; the inbound interface handles incoming traffic, i.e., results of remote requests that 'return' to this node and remote requests to access memory at this node.

Fig. 16.2 also shows a queue of ready threads; whenever the processor performs a context switch (i.e., switch from one thread to another), a thread from this queue is selected for execution and the execution continues until another context switch is performed.

Switching from one thread to another can take place:

(a) for each long–latency memory access [3] (a typical approach when context switching can be done very quickly),

(b) for each long–latency remote memory access [2] (a typical approach when the time of context switching is comparable with the memory cycle; in this case the processor is stalled while the thread accessing local memory waits for the result),

(c) after every instruction [14] (this approach is advantageous for eliminating data dependencies which slow–down the pipeline; since consecutive instructions are from different threads, they have no data
dependencies); typically the number of threads is equal to the number of pipeline stages, so no inter-instruction dependencies can stall
the pipeline.

For the first two cases, when a context switch is performed as a result
of long–latency memory access, the current thread becomes 'suspended',
its request is directed to the memory module (through the interconnecting network), and when the result of this request is received, the thread
becomes 'ready' again and joins the queue of ready threads waiting for
another execution phase on the processor.

For case (c), the thread, after issuing a long–latency memory access
request, becomes 'waiting' for the result of the requested operation.
If a waiting thread is selected for execution, its 'slot' simply remains
empty (i.e., no instruction is issued), which is equivalent to a single–
cycle pipeline stall. Since the threads issue their instructions one ofter
another, only a few processor cycles are lost during a long–latency operation of a single thread.

The average number of instructions executed between long–latency
operations (and context switches) is called the runlength of a thread,
ℓ_t, and is one of important modeling parameters. It is directly related
to the probability that an instruction requests a long–latency memory
operation.

Another important modeling parameter is the probability of long–
latency accesses to local, p_ℓ, (or remote, $p_r = 1 - p_\ell$) memory; as the
value of p_ℓ decreases (or p_r increases), the effects of communication overhead and congestion in the interconnecting network (and its switches)
become more pronounced; for p_ℓ close to 1, the nodes can be practically
considered in isolation.

The (average) number of available threads, n_t, is yet another basic
modeling parameter. For very small values of n_t, queueing effects can
be practically neglected, so the performance can be predicted by taking
into account only the delays of system's components. On the other
hand, for large values of n_t, the system can be considered in saturation,
which means that one of its components will be utilized in almost 100 %,
limiting the utilization of other components as well as the whole system.
Identification of this 'limiting' component (called the bottleneck) also
allows to estimate the performance of the system.

2.   MODELS

Petri nets have become a popular formalism for modeling systems
that exhibit parallel and concurrent activities [12, 11]. In timed nets

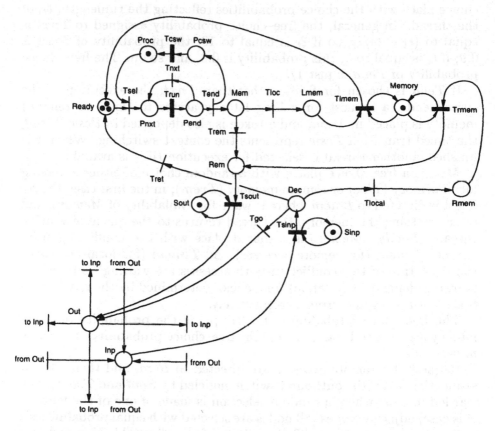

Figure 16.3 Instruction–level Petri net model of a multithreaded processor; case (a).

[16], deterministic or stochastic (exponentially distributed) firing times are associated with transitions, and transition firings occur in real–time, i.e., tokens are removed from input places at the beginning of the firing period, and they are deposited to the output places at the end of this period.

A timed Petri net model of a multithreaded processor at the level of instruction execution is shown in Fig.16.3.

The execution of each instruction of the 'running' thread is modeled by transition $Trun$. Place $Proc$ represents the (available) processor (if marked) and place $Ready$ – the queue of threads waiting for execution. The initial marking of $Ready$ represents the average number of available threads, n_t. It is assumed that this number does not change in time.

If the processor is available (i.e., $Proc$ is marked) and $Ready$ is not empty, a thread is selected for execution by firing the immediate transition $Tsel$. Execution of consecutive instructions of the selected thread is performed in the loop $Pnxt$, $Trun$, $Pend$ and $Tnxt$. $Pend$ is a free–

choice place with the choice probabilities reflecting the runlength, ℓ_t, of the thread. In general, the free–choice probability assigned to $Tnxt$ is equal to $(\ell_t - 1)/\ell_t$, so if ℓ_t is equal to 10, the probability of $Tnxt$ is 0.9; if ℓ_t is equal to 5, this probability is 0.8, and so on. The free–choice probability of $Tend$ is just $1/\ell_t$.

If $Tend$ is chosen for firing rather than $Tnxt$, the execution of the thread ends, a request for a long–latency access to (local or remote) memory is placed in Mem, and a token is also deposited in $Pcsw$. Firing the timed transition $Tcsw$ represents the context switching. When it is finished, another thread is selected for execution (if it is available).

Mem is a free–choice place, with a random choice of either accessing local memory ($Tloc$) or remote memory ($Trem$); in the first case, the request is directed to $Lmem$ where it waits for availability of $Memory$, and after accessing the memory, the thread returns to the queue of waiting threads, $Ready$. $Memory$ is a shared place with two conflicting transitions, $Trmem$ (for remote accesses) and $Tlmem$ (for local accesses); the resolution of this conflict (if both accesses are waiting) is based on marking–dependent (relative) frequencies determined by the numbers of tokens in $Lmem$ and $Rmem$, respectively.

The free–choice probability of $Tloc$, p_ℓ, is the probability of long–latency accesses to local memory; the free–choice probability of $Trem$ is $p_r = 1 - p_\ell$.

Requests for remote accesses are directed to Rem, and then, after a sequential delay (the outbound switch modeled by $Sout$ and $Tsout$), forwarded to Out, where a random selection is made of one of the four (in this case) adjacent nodes (all nodes are selected with equal probabilities). Similarly, the incoming traffic is collected from all neighboring nodes in Inp, and, after a sequential delay (the inbound switch $Sinp$ and $Tsinp$), forwarded to Dec. Dec is a free–choice place with three transitions sharing it: $Tret$, which represents the satisfied requests reaching their 'home' nodes; Tgo, which represents requests as well as responses forwarded to another node (another 'hop' in the interconnecting network); and $Tlocal$, which represents remote requests accessing the memory at the destination node. In the last case, the remote requests are queued in $Rmem$ and served by $Trmem$ when the memory module $Memory$ becomes available. The free–choice probabilities associated with $Tret$, Tgo and $Tlocal$ characterize the interconnecting network [8].

The traffic outgoing from a node (place Out) is composed of requests and responses forwarded to another node (transition Tgo), responses to requests from other nodes (transition $Trmem$) and remote memory requests originating in this node (transition $Trem$).

Instruction dependencies, within each thread, occasionally stall the pipeline to delay the execution of an instruction until its argument is available. Some dependencies can be removed by reordering the instructions or by renaming the registers (either by compiler or by hardware in

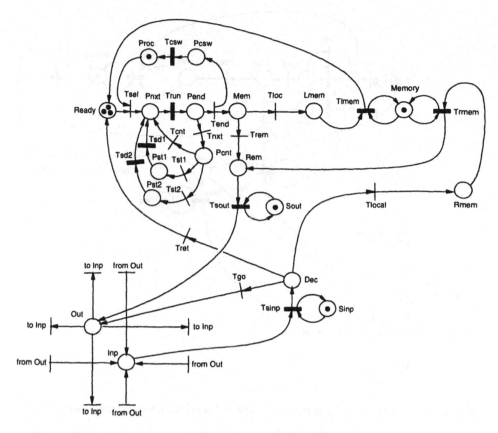

Figure 16.4 Instruction–level Petri net model of a multithreaded processor; case (a) with pipeline delays.

the instruction issue unit); the remaining dependencies are detected in the pipeline, and they stall the pipeline for one or more processor cycles. Pipeline stalls are not represented in Fig. ??.

Fig. 16.4 shows a modification of the model from Fig. 16.3 in which the transition $Tnxt$ is augmented by a free–choice place $Pcnt$ with three choices: $Tcnt$ which represents continuation without stalling, $Tst1$ which introduces a single–cycle stall (timed transition $Tsd1$), and $Tst2$ which introduces a two–cycle stall (timed transition $Tsd2$).

The choice probabilities associated with these three transitions characterize the frequency and the durations of pipeline stalls. In evaluation of this model, these choices are described by two probabilities, p_{s1} and p_{s2}, associated with $Tst1$ and $Tst2$, respectively (the probability associated with $Tcnt$ is simply $1 - p_{s1} - p_{s2}$). Although only two cases of pipeline stalls are represented in Fig. 16.4, other cases can be modeled in a similar way.

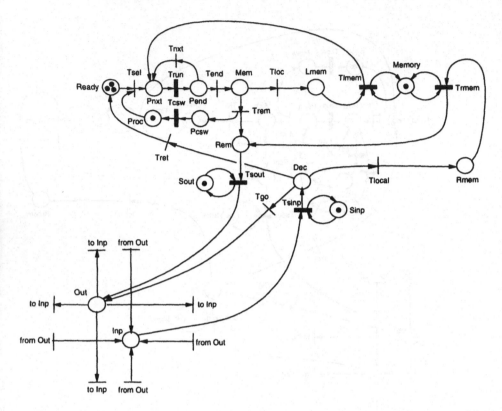

Figure 16.5   Instruction–level Petri net model of a multithreaded processor; case (b).

Fig. 16.5 shows a processor's model for the case when the context switching is performed for remote memory accesses only (case (b)). In this case, if the long–latency request is to local memory ($Tloc$), the processor 'waits' for the completion of the memory access and then continues the execution of the same thread. If the request is to remote memory ($Trem$), the processor performs context switch and executes another thread(s) concurrently with the remote memory access.

Fig. 16.5 does not represent the pipeline stalls; if needed, they can be added in the same way as in Fig. 16.4.

Petri net model of fine–grain multithreading (case (c)), in which consecutive instructions are (cyclically) issued from consecutive threads, is more elaborate because of the representation of the cyclic thread selection. An outline of the model is shown in Fig. 16.6, with a shaded area that models the switching from one thread to another (and suspending the threads during their long–latency operations). It should be observed that this outline is basically the same as in Figures 16.3, 16.4 and 16.5.

The detailed representation of the shaded area in Fig. 16.6 (with the adjacent elements), for a processor with four threads, is shown in Fig.

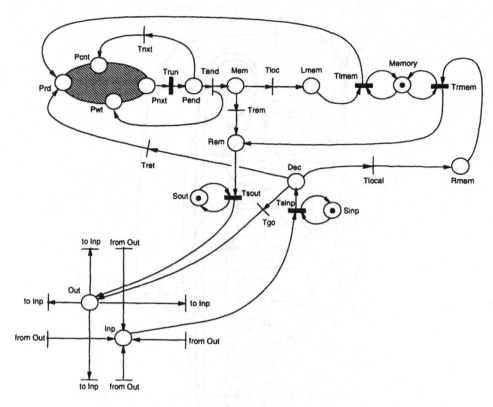

Figure 16.6   Instruction–level Petri net model of a multithreaded processor; case (c).

16.7. Fig. 16.7 may seem to be a bit complicated, but it has a regular structure that is repeated for each of the four threads (and would also be used for additional threads).

This basic structure, for thread "2", is shown in Fig. 16.8. The idea of this model is as follows. If the thread is "ready", a token is waiting in $Pth2$ for a "control token" to appear in $Ps2$ (the marking of $Ps2$ in Fig. 16.8 indicates that an instruction from thread "2" is going to be issued in the next processor cycle). Place $Ps2$ is an element of a "thread ring" (in Fig. 16.7 this ring connects $Ps1$, $Ps2$, $Ps3$, $Ps4$ and back to $Ps1$, and there are several different ways connecting consecutive threads). This "thread ring" contains a single token ($Ps1$ in Fig.2.5 and $Ps2$ in Fig. 16.8).

If the selected thread is "ready", the firing of $Tth2$ inserts a token in $Pnxt$ (the next instruction to be executed by $Trun$), and also a token in $Pr2$. If the issued instruction does not perform long–latency memory access, the free–choice transition $Tnxt$ is fired (with the probability depending upon the thread runlength ℓ_t), and a token is deposited in $Pcnt$.

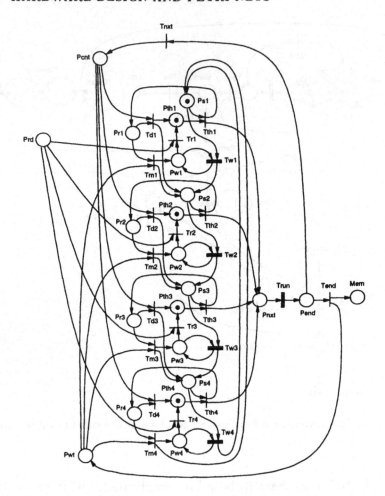

Figure 16.7 Thread switching in a fine-grain multithreaded processor.

This token (together with a token in $Pr2$) enables firing of $Td2$, which regenerates a token in $Pth2$ and forwards the control token to $Ps3$.

If transition $Tend$ is selected for firing rather than $Tnxt$, a long–latency memory access (local or remote) is initiated, and a token is deposited in Pwt. In this case $Tm2$ is enabled to fire, which inserts a token in $Pw2$ (to indicate that the thread is waiting for termination of its long-latency memory access), and also the control token is forwarded to $Ps3$.

If a thread is "waiting" and a selection token appears in $Ps2$, the timed transition $Tw2$ fires and, after a unit of time (one processor cycle), deposits a control token in $Ps3$.

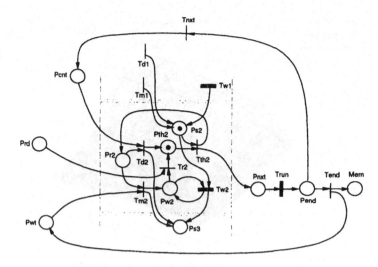

Figure 16.8 Single thread switching section.

Finally, when the long–latency memory operation (local or remote) is completed, a token is deposited in Prd (Fig. 16.6), and the "waiting" thread becomes "ready" by firing $Tr2$.

3. PERFORMANCE

It is convenient to assume that all timing characteristics are expressed in processor cycles (which is assumed to be 1 unit of time). The basic model parameters and their values used in subsequent evaluations are as follows:

symbol	parameter	values
n_t	the (average) number of threads	2,...,20
ℓ_t	thread runlength	5,10,20
t_{cs}	context switching time	1,2,5
t_m	memory cycle time	10
t_s	switch delay	5,10
p_ℓ, p_r	probability of accesses to local/remote memory	0.1,...,0.9
p_{s1}, p_{s2}	probabilities of pipeline stalls	0.1,0.2

Fig. 16.9 shows the utilization of the processor as a function of the number of available threads, n_t, and the probability of long–latency accesses to local memory, p_ℓ, for fixed values of other modeling parameters. It can be observed that, for values of p_ℓ close to 1, the utilization increases with the number of available threads n_t, and tends to the bound

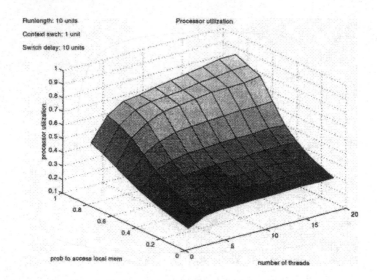

Figure 16.9   Processor utilization; $t_{cs} = 1$, $\ell_t = 10$, $t_s = 10$.

0.91 which is determined, in this case, by the ratio of $\ell_t/(\ell_t + t_{cs})$ (the context switching time, t_{cs}, is an overhead of multithreading).

For smaller values of p_ℓ, the utilization of the processor 'saturates' very quickly and is practically insensitive to the number of available threads n_t. This is a clear indication that some other component of the system is the bottleneck.

The bottlenecks can be identified by comparing service demands for the different components and the system [9]; the component with the highest service demand is the first to reach its 'saturation' (i.e., utilization of almost 100%) which limits the performance of all other elements of the system.

The service demands (per one long–latency memory access) are [17]:

component	service demand
processor	ℓ_t
memory	t_m
inbound switch	$2 * p_r * n_h * t_s$
outbound switch	$2 * p_r * t_s$

where n_h is the average number of hops (in the interconnecting network) that a request must perform to reach its destination (for a 16–processor system, with a uniform distribution of accesses over the nodes, the value of n_h is close to 2 [8]; in general, for a system with $p \times p$ processors connected by a 2–dimensional torus network, n_h can be approximated reasonably well by $p/2$).

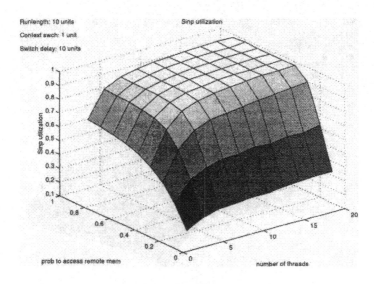

Figure 16.10 Inbound switch utilization; $t_{cs} = 1$, $\ell_t = 10$, $t_s = 10$.

If $\ell_t = t_m = t_s$, and $n_h = 2$, the inbound switch becomes the bottle-neck for $p_r > 0.25$; for $p_r < 0.25$ the processor and the memory are the bottlenecks (their service demands are equal for $t_m = \ell_t$).

Fig. 16.10 shows the utilization of the inbound switch (for the same values of modeling parameters as in Fig. 16.9); it should be noted that in Fig. 16.10 (as well as in Fig. 16.11) the probability of accessing remote memory, p_r, is used instead of p_ℓ, so the 'front part' of Fig. 16.10 corresponds to the 'back part' of Fig. 16.9 and vice versa.

Fig. 16.10 shows that the inbound switch enters its saturation quite quickly for increasing values of n_t and p_ℓ. The delay introduced by the inbound switch is simply too large if the probability of accesses to remote memory, p_r, can be greater than 0.25.

Fig. 16.11 shows the utilization of the inbound switch for the switch delay $t_s = 5$. In this case the 'saturated' region is much smaller than in Fig. 16.10. The corresponding utilization of the processor is shown in Fig. 16.12; the utilization is significantly better than in Fig. 16.9, but the limiting effects of the inbound switch can still be observed for small values of p_ℓ.

The influence of pipeline stalls is rather straightforward to predict; since the stalled processor cycles are lost, the performance of the processor must decrease when pipeline stalls are taken into account.

Figures 16.13 and 16.14 show the utilization of the processor (as a function of the number of threads, n_t, and the probability of accesses to local memory, p_ℓ, as before), for two different probabilities of pipeline stalls. Fig. 16.13 corresponds to the case when $p_{s1} = p_{s2} = 0.1$, i.e.,

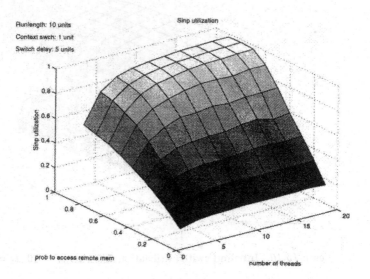

Figure 16.11   Inbound switch utilization; $t_{cs} = 1$, $\ell_t = 10$, $t_s = 5$.

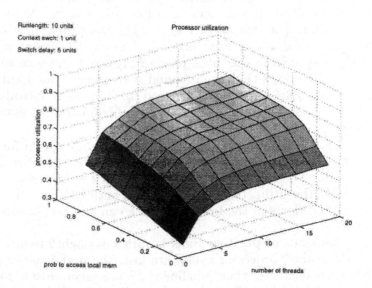

Figure 16.12   Processor utilization; $t_{cs} = 1$, $\ell_t = 10$, $t_s = 5$.

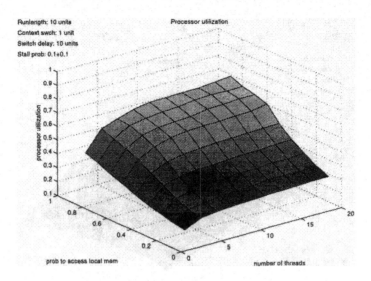

Runlength: 10 units
Context swch: 1 unit
Switch delay: 10 units
Stall prob: 0.1+0.1

Figure 16.13 Processor utilization; $t_{cs} = 1$, $\ell_t = 10$, $t_s = 10$, $p_{s1} = p_{s2} = 0.1$.

for 10% of instructions the pipeline stalls for one cycle, and for another 10% of instructions the pipeline is stalled for two processor cycles. Consequently, the processor is stalled approximately 25% of time, so its utilization is expected to be approximately 25% lower than in the case without pipeline delays (Fig. 16.9). Indeed, a comparison of Fig. 16.13 and Fig. 16.9 shows such a difference for the values of p_ℓ close to 1. For $p_\ell < 0.75$ (or $p_r > 0.25$), the inbound switch is the bottleneck, so there is no significant difference between Fig. 16.13 and Fig. 16.9 because in both cases the performance is determined by the delay of the inbound switch (which is the same for both cases).

Fig. 16.14 shows the utilization of the processor which is stalled approximately 40% of time ($p_{s1} = p_{s2} = 0.2$). It should be noted that, for the values of p_ℓ close to 1, the utilization is further reduced, as expected, while the utilization in the "saturated region" (i.e., for small values of p_ℓ) is not affected by the pipeline stalls, and remains the same as in Figures 16.13 and 16.9.

The influence of the context switching time on the processor utilization is shown in Figures 16.15 and 16.16. Because the time of context switching is an overhead for the thread execution, the upper bound on the processor utilization is $\ell_t/(\ell_t + t_{cs})$. For $\ell_t = 10$ and $t_{cs} = 1$ (Fig. 16.9), this bound is 0.91. For $\ell_t = 10$ and $t_{cs} = 2$ (Fig. 16.15), this bound becomes 0.83, and for $\ell_t = 10$ and $t_{cs} = 5$ (Fig. 16.16), the upper bound decreases to 0.67. The influence of these upper bounds can easily be observed in Figures 16.15 and 16.16 for the values of p_ℓ close to 1.

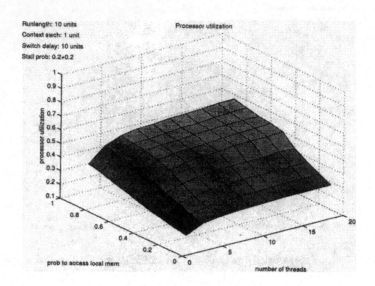

Figure 16.14   Processor utilization; $t_{cs} = 1$, $\ell_t = 10$, $t_s = 10$, $p_{s1} = p_{s2} = 0.2$.

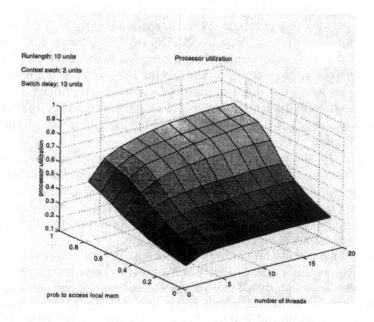

Figure 16.15   Processor utilization; $t_{cs} = 2$, $\ell_t = 10$, $t_s = 10$.

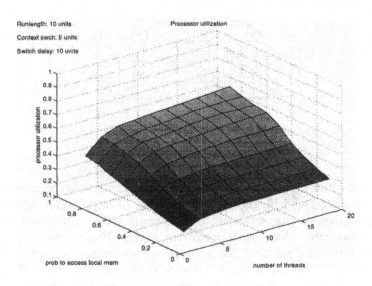

Figure 16.16 Processor utilization; $t_{cs} = 5$, $\ell_t = 10$, $t_s = 10$.

Similarly to the probability of pipeline stalls, the context switching time has little effect on processor utilization when the performance is limited by the inbound switch (i.e., for small values of p_ℓ).

The combined effect of pipeline stalls and the context switching time is not difficult to predict. Since both factors reduce the number of processor cycles used for execution of program instructions, their effects are cumulative with respect to reducing the utilization of the processor.

If context switching is performed for long–latency remote memory accesses (case (b)), the context switching time t_{cs} (comparable to t_m in this case) introduces a considerable overhead, so the upper bound on the utilization of the processor, for $t_{cs} = t_m$, is $\ell_t/(\ell_t + t_{cs})$; for $t_{cs} = t_m = \ell_t$, this upper bound is equal to 0.5; for $\ell_t = 20$ and $t_{cs} = t_m = 10$, the bound increases to 0.67. Consequently, much lower values of processor utilization are obtained in this case; to achieve a reasonable performance, large values of the runlength, ℓ_t, are needed.

4.   CONCLUDING REMARKS

All Petri net models of multithreaded distributed memory architectures discussed in this paper have a common part that represents the memory and the interconnecting network; the differences between models are in the context switching part and in the nature of context switching. Fig. 16.6 shows the general 'framework' of models with the shaded area either representing case (c), as in Fig. 16.7, or case (a) which could easily be extracted form Fig. 16.3 or Fig. 16.4; a few more model elements should be included in the shaded area to also cover case (b).

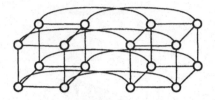

Figure 16.17   Outline of a 16–processor system.

For performance analysis of derived models, the interconnecting network is characterized by the average number of hops, n_h. Consequently, different networks characterized by the same value of n_h will yield the same performance characteristics of the nodes. For example, Fig. 16.17 shows a hypercube network for a 16–processor system that can be used instead of a 2–dimension torus network shown in Fig. 16.1. Since each node in Fig. 16.17 is connected to 4 neighbors (as is the case in Fig. 16.1), the average numbers of hops (with the same assumptions as before) for the two networks are the same, and then the performance characteristics for the two types of interconnecting networks are identical.

Moreover, models of systems with different numbers of processors (e.g., 25, 36, etc.) require only minor adjustment of a few model parameters (the free–choice probabilities describing the traffic of messages in the interconnecting network); otherwise the models are as presented in this paper.

One of the assumptions of the derived models was that accesses to memory are uniformly distributed over the nodes of the system. If this assumption is not realistic and some sort of 'locality' is present, the only change that needs to be made is an adjustment of the value of n_h; for example, if the probability of accessing nodes decreases with the distance (i.e., nodes which are close are more likely to be accessed that the distant ones), the value of n_h will be smaller than that determined for the uniform distribution of accesses, and will result in improved performance (especially for values of p_r close to 1).

In many cases the presented models can be abstracted to a less detailed form without any significant loss of accuracy. One of such transformations replaces the instruction–level model of a processor by its thread–level model; Fig. 16.18 shows such a transformation applied to the model shown in Fig. 16.3. In Fig. 16.18 the firing time of *Trun* is exponentially distributed with the average value equal to the thread runlength; it represents the (random) execution time of a thread (with no individual instructions). It appears that this simplification does not affect the results in any significant way, but it speeds up (several times) the simulation of the model, eliminating many events which do not actually contribute to the performance characteristics.

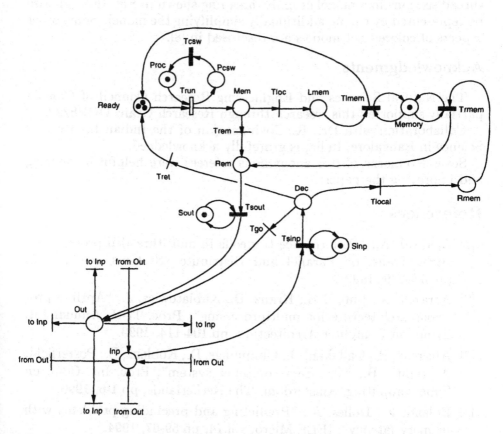

Figure 16.18   Thread–level Petri net model of a multithreaded processor, case (a).

Petri net models of multiprocessor systems contain many 'regularities' which can be used for model reduction in a more sophisticated modeling formalism. For example, in colored Petri nets [10], tokens are associated with attributes (called colors), so different activities can be associated with tokens of different types. An immediate application of colors is to represent the different processors (or nodes) by different colors within the same processor model; consequently, a colored Petri net will need only one processor model (for any number of processors). Similarly, the thread sections in a model of multithreading shown in Fig. 16.7 can also be represented by colors, additionally simplifying the model. Some other aspects of colored net models are discussed in [8].

Acknowledgments

The Natural Sciences and Engineering Research Council of Canada partially supported this research through Research Grant OGP8222.

Collaboration with Dr. R. Govindarajan of the Indian Institute of Science in Bangalore, India, is gratefully acknowledged.

Several remarks of two anonymous referees were helpful in revising and improving the paper.

References

[1] Agarwal, A., "Performance tradeoffs in multithreaded processors"; IEEE Trans. on Parallel and Distributed Systems, vol.3, no.5, pp.525-539, 1992.

[2] Agrawal, A., Lim, B-H., Kranz, D., Kubiatowicz, J., "April: a processor architecture for multiprocessing"; Proc. 17-th Annual Int. Symp. on Computer Architecture, pp.104-114, 1990.

[3] Alverson, R., Callahan, D., Cummings, D., Koblenz, B., Posterfield, A., Smith, B., "The Tera computer system"; Proc. Int. Conf. on Supercomputing, Amsterdam, The Netherlands, pp.1-6, 1990.

[4] Boland, K., Dolles, A., "Predicting and precluding problems with memory latency"; IEEE Micro, vol.14, pp.59-67, 1994.

[5] Boothe, B., Ranade, A., "Improved multithreading techniques for hiding communication latency in multiprocessors"; Proc. 19-th Annual Int. Symp. on Computer Architecture, pp.214-223, 1992.

[6] Byrd, G.T., Holliday, M.A., "Multithreaded processor architecture"; IEEE Spectrum, vol.32, no.8, pp.38-46, 1995.

[7] Govindarajan, R., Nemawarkar, S.S., LeNir, P., "Design and performance evaluation of a multithreaded architecture"; Proc. First IEEE Symp. on High–Performance Computer Architecture, Raleigh, NC, pp.298-307, 1995.

[8] Govindarajan, R., Suciu, F., Zuberek, W.M., "Timed Petri net models of multithreaded multiprocessor architectures"; Proc. 7-th Int. Workshop on Petri Nets and Performance Models (PNPM'97), St. Malo, France, pp.153-162, 1997.

[9] Jain, R., "The art of computer systems performance analysis"; J. Wiley & Sons 1991.

[10] Jensen, K., "Coloured Petri nets"; in: "Advanced Course on Petri Nets 1986" (Lecture Notes in Computer Science 254), Rozenberg, G. (ed.), pp.248-299, Springer–Verlag 1987.

[11] Murata, T., "Petri nets: properties, analysis and applications"; Proceedings of IEEE, vol.77, no.4, pp.541–580, 1989.

[12] Reisig, W., "Petri nets - an introduction" (EATCS Monographs on Theoretical Computer Science 4); Springer–Verlag 1985.

[13] Sinharoy, B., "Optimized thread creation for processor multithreading"; Computer Journal, vol.40, no.6, pp.388-399, 1997.

[14] Smith, B.J., "Architecture and applications of the HEP multiprocessor computer System"; Proc. SPIE – Real-Time Signal Processing IV, vol. 298, pp. 241–248, 1981.

[15] Weber, W.D., Gupta, A., "Exploring the benefits of multiple contexts in a multiprocessor architecture: preliminary results"; Proc. 16-th Annual Int. Symp. on Computer Architecture, pp.273-280, 1989.

[16] Zuberek, W.M., "Timed Petri nets – definitions, properties and applications"; Microelectronics and Reliability (Special Issue on Petri Nets and Related Graph Models), vol.31, no.4, pp.627–644, 1991.

[17] Zuberek, W.M., Govindarajan, R., "Performance balancing in multithreaded multiprocessor architectures"; Proc. 4-th Australasian Conf. on Parallel and Real–Time Systems (PART'97), Newcastle, Australia, pp.15-26, 1997.